# BIRD MIGRATION

# BIRD MIGRATION

DOMINIC COUZENS

First published in 2005 by New Holland Publishers (UK) Ltd
London • Cape Town • Sydney • Auckland

Garfield House, 86–88 Edgware Road, London W2 2EA, United Kingdom
www.newhollandpublishers.com

80 McKenzie Street, Cape Town, 8001, South Africa

Level 1/Unit 4, 14 Aquatic Drive, Frenchs Forest, NSW 2086, Australia

218 Lake Road, Northcote, Auckland, New Zealand

10 9 8 7 6 5 4 3 2 1



ISBN 1 84330 970 X

Publishing Manager: Jo Hemmings
Project Editor: Charlotte Judet
Design: D & N Publishing, Hungerford, Berkshire
Cartographer: William Smuts
Production: Joan Woodroffe

Reproduction by Modern Age Repro Co., Hong Kong
Printed and bound in Malaysia by Times Offset (M) Sdn Bhd



Front jacket, Barnacle Geese; Back jacket, White-winged Black Terns.

# THE WILDLIFE TRUSTS

The Wildlife Trusts partnership is the UK's leading voluntary organization working, since 1912, in all areas of nature conservation. We are fortunate to have the support of more than 530,000 members, including some famous household names.

The Wildlife Trusts protect wildlife for the future by managing in excess of 2,500 nature reserves, ranging from woodlands and peat bogs, to heathlands, coastal habitats and wildflower meadows. We campaign tirelessly on behalf of wildlife, including of course the multitude of bird species.

We run thousands of events, including dawn chorus walks and birdwatching activities, and projects for adults and children across the UK. Leicestershire and Rutland Wildlife Trust organizes the British Birdwatching Fair at Rutland Water – now also home to Osprey. The Wildlife Trusts work to influence industry and government and also advise landowners.

As numbers of formerly common species plummet we are urging people from all walks of life to take action, whether supporting conservation organizations in their work for birds, or taking a few small steps such as providing food and water for garden birds.

The Wildife Trusts manage some of the most important sites in the UK for birds. Whether it is Puffins on Skomer Island, Ospreys at Loch of the Lowes in Scotland and Rutland Water, or Bitterns at Far Ings in Lincolnshire, Wildlife Trust reserves offer fantastic birdwatching opportunities.

For many people bird migration is one of the most fascinating aspects of bird activity, but few of us really understands what drives birds to travel such long distances in inhospitable conditions, or how they know exactly in which direction to fly. *Bird Migration* is an excellent introduction to the fantastic feat of endurance that is bird migration. It describes all the different types of migration, explains how birds prepare for their journeys, and unravels the mysteries of the orientation progammes that allow birds to fly from A to B and back again. Dominic Couzen's easy-to-follow text also describes the relationship between weather and migration and discusses ways to study and observe migration. To complete the book there are descriptions of the migration journeys of six of the most amazing migrant species and a comprehensive chart of all bird species that migrate to or from the UK. This shows whether the 240-plus species are residents or summer or winter visitors, when they arrive and depart, their type of migration, and more. *Bird Migration* is a fascinating and highly enjoyable book that will help answer all questions regarding this most remarkable of bird behaviours.

The Wildlife Trusts is a registered charity (number 207238). For membership, and other details, please phone The Wildlife Trusts on 0870 0367711, log on to www.wildlifetrusts.org or complete the form on the inside back cover.

# CONTENTS

# INTRODUCTION

Of all the many and varied types of bird behaviour, few catch the imagination of both the enthusiast and general public more than the phenomenon we call migration. In part its thrill is in its easy appreciation – anyone can witness bird movements, and each one of us can delight in hearing the first Cuckoo or seeing the first Swallow of spring. The other part of the thrill is in migration's mystery, for although we know that birds can travel vast distances with mind-boggling accuracy, we still don't understand exactly how they do it. Certainly, migration is far beyond our own abilities – at least, without the help of machines or navigational aids – and that makes us ask questions.

As it turns out – a lot of questions. Few zoological subjects receive more attention and study than bird migration. More than 300 papers – and sometimes a great deal more – are published every year on migration, and that amounts to a huge mass of information. Much of it is very specific and highly technical, and it covers many different disciplines: zoology, physics, biochemistry, mathematics, and so on. Each year a few discoveries capture the imagination, such as the recent finding that racing pigeons like to follow roads instead of flying on a straight path to their lofts, but many studies remain buried under the mountain of literature, even though they have important things to say. So any book about migration has to draw a lot of strands together, which must weave into a reasonably coherent whole, or theme.

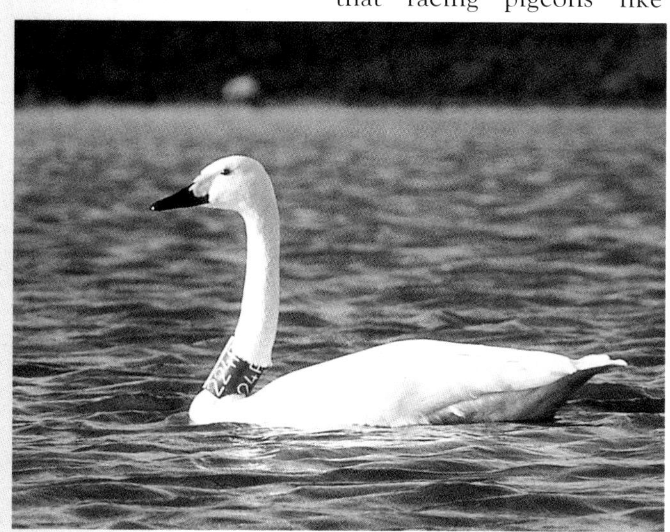

*Bewick's Swan with neck collar for tracking purposes.*

Not only does the book's theme give it direction, it also, mercifully, dictates what can be left out.To that end, this book takes very much a birdwatcher's view of migration. Although it aims to cover all the main aspects, it focuses on what a birder might see and experience and question. A birdwatcher might ask, for example, where a bird is going, and where it has come from. An observer might want to find out how a bird manages to get somewhere, and how it knows when and where to go. A birdwatcher may want to know where to see migrating birds or observe ringing, or how to interpret the weather in terms of its impact on migration. As a result, this book includes as many practical tips and suggestions as possible, in addition to scientific findings. Throughout, I have tried to maintain the perspective of someone who wishes to understand migration when they go into the field. It should be stated from the outset that this book is intended for birdwatchers from Britain and north-west Europe. Most, although not all, the birds mentioned come from this region.

*The Common Tern. Migrating birds can be seen offshore in autumn.*

It is also written from the point of view of what could be described as a 'knowledgeable beginner'. This is important, because there are plenty of more technical books. In this regard I can recommend three important books that anyone interested in migration ought to read, perhaps after this one. I have drawn on all three in preparing the book you are holding now. They are: Peter Berthold's Bird Migration: A General Survey (2nd edition); Thomas Alerstam's Bird Migration; and the monumental Migration Atlas: Movements of the Birds of Britain and Ireland, recently published by the British Trust for Ornithology (BTO). Perhaps the success of this present title can be measured partly by how many people are inspired to pick up the pages of these more learned works.

*Flock of male Pochards. British breeding birds disperse in winter, some to Europe, and many birds move here in winter from northern and eastern Europe.*

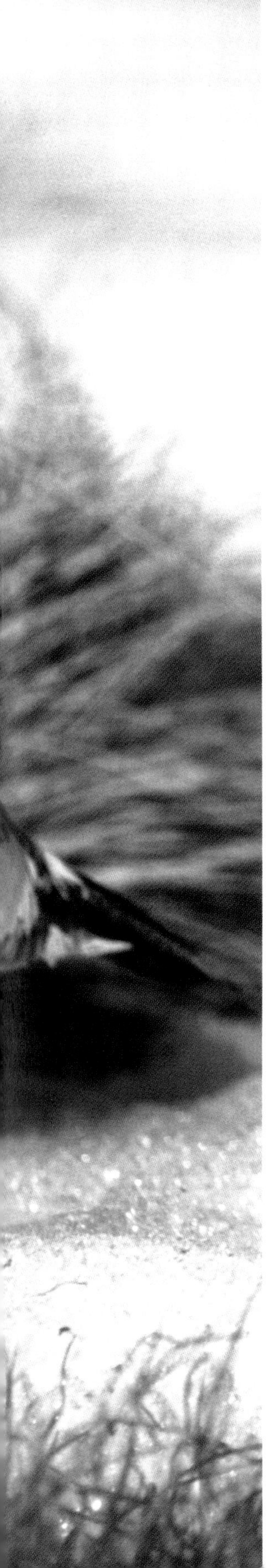

# 1
# BIRD MOVEMENTS AND TYPES OF MIGRATION

*What is Migration?*
*It might be the opening question to this book, but an all-embracing answer is surprisingly elusive because there is no clear, recognized definition for migration. Birds undertake a great variety of different movements, and some of them are easily understood as migrations and some are not.*

The Ptarmigan – so is it a migrant?

## Defining Migration

The widely held perception of migration is too narrow. We all know that Swallows depart Britain for the winter and return in the spring, having travelled many thousands of kilometres in between. But that's only one end of a complicated scale. What about the other end? In late summer, juvenile birds of almost every species leave their parents' territory and wander around the neighbourhood, perhaps eventually settling less than a kilometre away. It is a definite movement, with a purpose – but is it a migration?

**Daily Commute**

*Northern Gannets provide a good example of the difficulty of defining migration. During the breeding season adults may make daily fishing trips of over 300km to collect food for their young – further than the distance of many birds' complete migration!*

You can make up your own mind as to what is and is not 'migration' by reading the sections that follow. But, in terms of coverage, this book will take the broadest view – that is, it will embrace any movement of all or part of a species' population from one place to another, with a purpose. The only movements to be left out completely are those that take place regularly within the course of a day: a movement between a feeding site and a roost site, for example. These are definitely not migrations.

Some terms will be used throughout this book. A bird that migrates is called a **migrant**. If it regularly migrates to and fro between one place and another, we will label the two ends of its journey as the **breeding grounds** and the **wintering grounds**, respectively. Birds that do not migrate (although their young might disperse) are called **resident** or **sedentary**. Birds that pass by a place on migration, but do not breed or spend the winter at the location concerned, are called **passage migrants**. The movement of any bird passing through a site is often known as 'passage', as in the following example: 'We've had an impressive passage of terns recently'. However, I will avoid such usage of the word here.

## To and Fro Migration

This is the type of migration that most of us understand best. A bird breeds in one place, spends the winter in a different place, often far away, and then returns in the spring, often to exactly the same spot where it bred the previous year The phenomenon of returning to the same breeding area year after year is known as **breeding site fidelity**. After breeding, it will leave once again to follow the same route, often to the same wintering spot. This is known as **wintering site fidelity**. And so on, to and fro.

A migrant's travels are essential for survival. Many species in northern Europe, for example, must leave the breeding site in autumn because the winters there are too cold and there will not

be enough food to eat. Remaining at the same site would be fatal. The wintering site, however, is a warm and comparatively food-rich refuge. But there will be a lot of competition for food in such a location, and it will pay the migrant to return north when conditions are inviting once again. This is what many familiar birds do, including Swallows and Willow Warblers. For the rest of this book, we will use the term **outward migration** for the journey a bird undertakes after breeding, and **return migration** for the journey a bird makes back to its breeding site.

However, a bird's migratory behaviour is seldom so straightforward. There are many variations on the theme of to and fro migration, leading to many different strategies and patterns. For example, although many species travel to and fro, flying out one way and retracing their steps on the way back, a good number use a different route for their return migration – much as drivers might use the M1 to go south and then the M6 to go north. In Britain, birdwatchers see plenty of Little Stints and Curlew Sandpipers on muddy lagoons in the autumn during the birds' outward migration, but very few in the spring; this is because, in the spring, these waders return much further to the east of the UK, taking a direct route north to their breeding sites. Such a pattern is

*Willow Warbler (top) and Chiffchaff (bottom), two very similar species with different migrations. The longer-winged Willow Warbler travels much further than the shorter-winged Chiffchaff.*

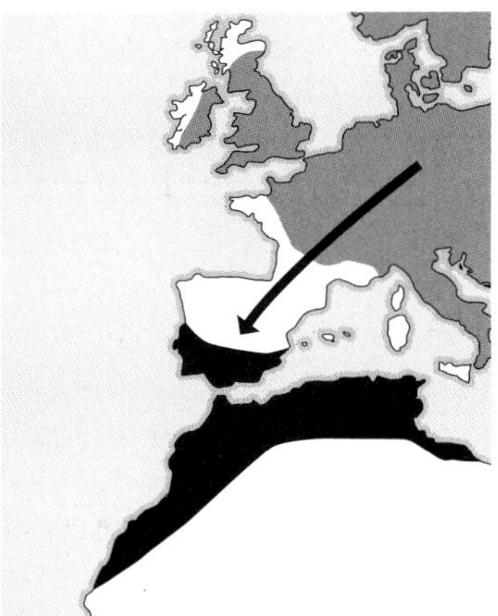

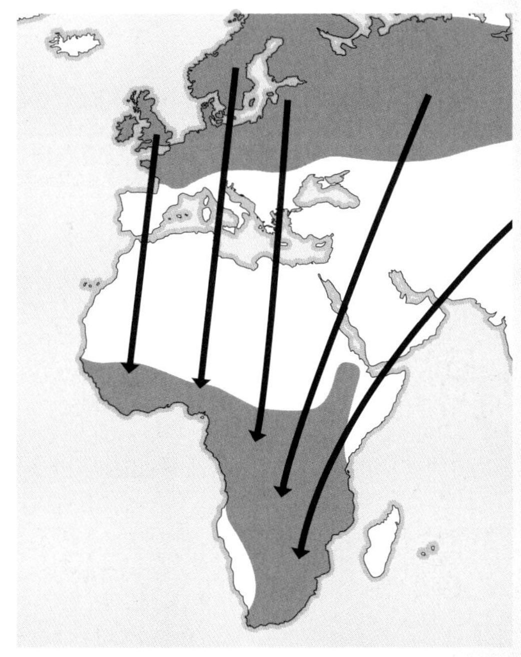

(Above) *Migratory journeys vary in length. Blackcaps travel a relatively short distance, whereas the migration of the Willow Warbler* (Right), *a typical to and fro migrant, is considerably further. (top, breeding areas; bottom, wintering areas).*

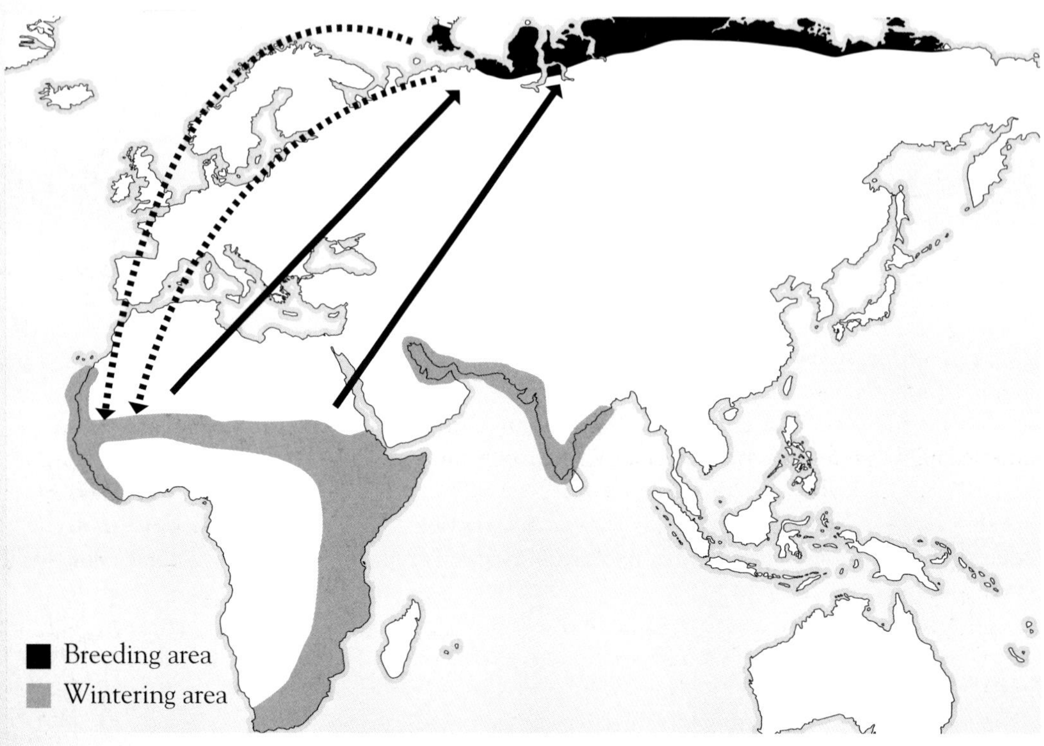


*Map of migratory path of Little Stint (below), a classic Loop Migrant. From its wintering grounds in Africa it takes a direct route north in spring (thick line) on its return migration, but takes a more westerly route outward (dotted line) in autumn.*

known as **loop migration**, because, if you use your imagination, the path describes a loop. Even more impressive loop migrations are performed by some seabirds, such as the Great Shearwater (*see* map on page 119).

Whichever way a bird is going, there is no guarantee that it will fly in a straight line. Birds often follow useful geographical features such as coastlines and rivers. And they may need to take a detour to avoid some geographical hazard such as a mountain range, or to take advantage of local topography such as a narrow sea crossing. We might undertake the same sort of detour on our journeys, perhaps to avoid going directly through a city centre, or to find our way to a bridge across a river. If we plot the route that some migrants take, it is clear that their migratory path has a bend or hook in it. This is called a **hooked migration** or **arched migration**, the latter term catering for the more gradual deviations from a straight course.

If the geographical hazard is large and the breeding distribution of a certain species is wide, it might be worthwhile for birds from one area to go one side of the obstacle, and others to go to the other side. For instance, in Europe the Mediterranean Sea is a major barrier, and one might expect birds from western Europe to

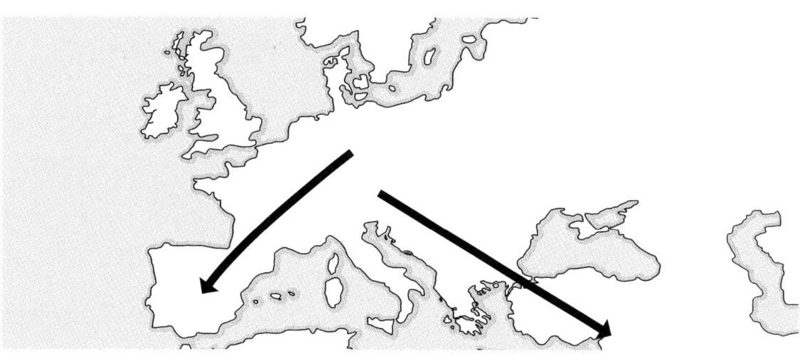

*The migratory divide of the Spotted Flycatcher (above and left). Birds breeding west of 12°E of longitude initially head south-west on their outward migration, and those to the east initially orient south-east.*

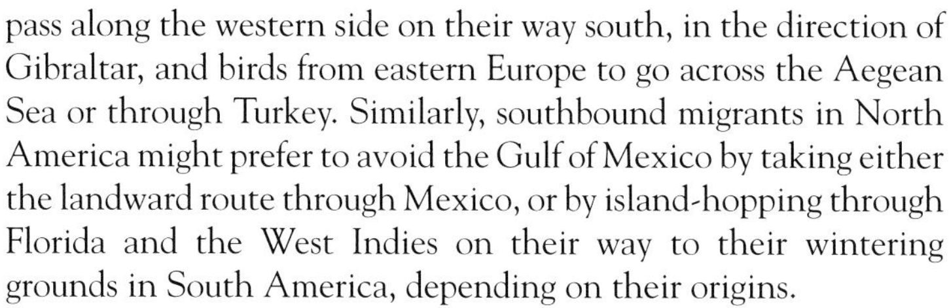

pass along the western side on their way south, in the direction of Gibraltar, and birds from eastern Europe to go across the Aegean Sea or through Turkey. Similarly, southbound migrants in North America might prefer to avoid the Gulf of Mexico by taking either the landward route through Mexico, or by island-hopping through Florida and the West Indies on their way to their wintering grounds in South America, depending on their origins.

This, indeed, is what many migrants do, and such patterns of divergent movement are called **migratory divides**. A good example in Europe is the Spotted Flycatcher: birds breeding west of 12° E of longitude orient initially to the south-west on their outward migration, while birds that breed to the east of this line head to the south-east. There is a small area of overlap in which birds might travel in either direction, but either side of this zone very few, if any, birds take the 'wrong' route.

## Many Routes

Although many birds take a detour to avoid large barriers on their migration, this does not necessarily mean that the whole travelling population will be condensed into a small flyway, a sort of aerial motorway. The majority of migrants, including those that avoid barriers as well as those that don't, actually travel on what is known as a **broad front**. One could imagine this, not as a motorway in the sky, but more as a series of very broad streams – perhaps 100km wide – all moving parallel along the same bearing (*see* diagram overleaf). The result is that populations of birds travelling south-west from France will not normally run into birds travelling south-west from Poland, but will fly parallel to them. And, on a smaller scale, birds from Kent will fly parallel to birds from Dorset.

The opposite of a broad front is, of course, a **narrow front**. Relatively few species practise this kind of migration and they are mostly large, daytime migrants such as cranes and storks. When migrating these birds really do fly along narrow 'corridors' in the sky, and seldom stray off course.

*Schematic representation of broad front migration, a series of streams of birds all travelling on the same bearing, but parallel to each other.*

*Schematic representation of narrow front migration. The birds travel along narrow corridors between breeding and wintering areas, rarely straying off route. The wintering grounds are shaded.*

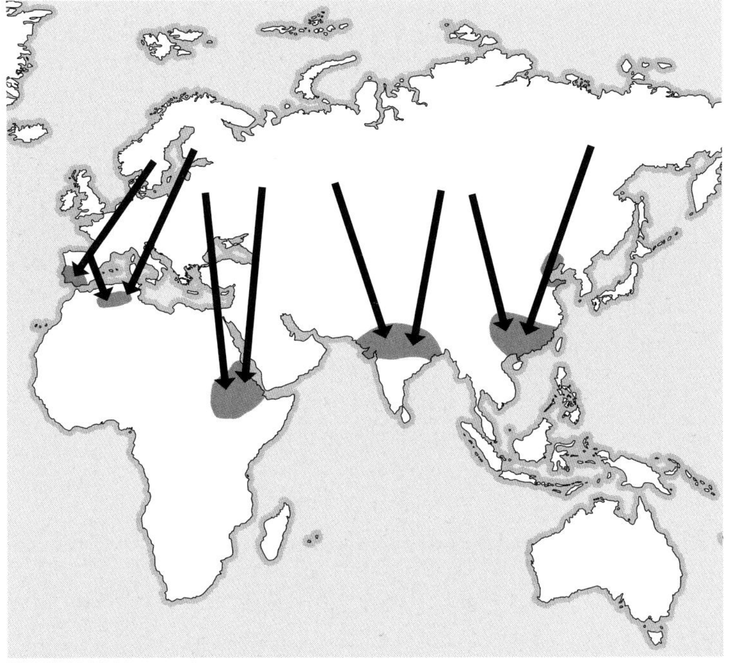

**A Very Unusual Migration**

*The Quail is thought to do something that no other European bird does, except perhaps the Spanish Sparrow: it migrates between broods. Having wintered in Africa, it migrates to southern Europe and breeds there in the spring. Then it continues its northward journey and settles in Britain, for example, where it raises a second brood before returning south.*

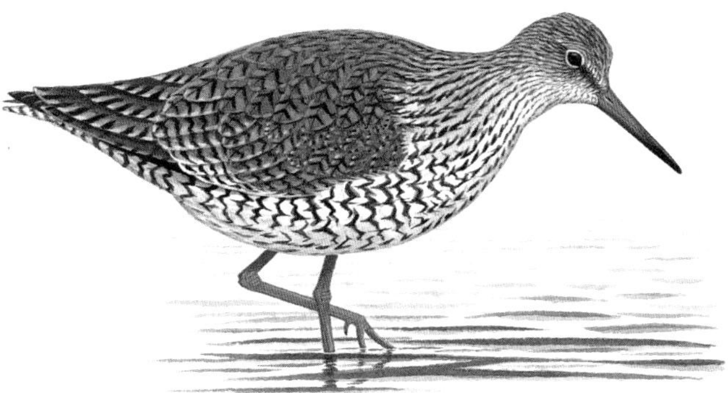

*Redshanks breeding in Norway and Sweden winter farther south than British Redshanks, an example of leapfrog migration.*

Finally, some birds exhibit an interesting phenomenon known as **leapfrog migration**, which in its broadest form occurs when one population of a species migrates further than another, and may catch it up and overtake it. In the relatively mild climate of north-west Europe, for example, a bird population may be resident and stay put during the winter, while at the same time more northern populations of the species may evacuate their cooler breeding grounds and fly straight over these stay-at-homes on their way to wintering areas further south.

## Altitudinal Migration

Strictly speaking, **altitudinal migration** – a seasonal movement in altitude rather than latitude – is another form of to and fro migration. It shares the same inherent generating force: the need to shift location in order to escape inhospitable conditions. But it differs in so many ways from 'normal' migration that it deserves its own section here.

The most striking and distinctive feature of altitudinal migration is in the short distances covered. In a few cases, birds simply come down from the top of mountains into valleys, thereby covering as little as a kilometre. Ptarmigans are an extreme example of such 'migrants' – usually found at 1,200m or more in Scotland, they may be forced to descend to about 500m if conditions demand. This is a remarkably short journey in terms of both altitude lost and distance travelled, but it is sufficient for survival.

A more typical example of an altitudinal migrant in Europe and Asia is the Wallcreeper. In the breeding season this thrilling and charismatic species is found on precipitous cliff-faces in high mountains up to about 2,700m, but in the autumn it comes down to well below 1,000m and may disperse in all directions, travelling up to 50km away from its breeding areas and sometimes swapping its wild rock faces for the sides of buildings in towns. While snow falls and storms rage in the mountains, the Wallcreeper resides in the gentler climes of the lowlands.

It is also worth mentioning the highly unusual migration of the Water Pipit. This is another species that breeds in the high mountains of Europe, from Spain to the Carpathians. However, instead of fanning out and remaining close to its breeding sites, the Water Pipit migrates a considerable distance 'the wrong way' north or north-west in the autumn, to spend the winter in freshwater meadows and wetlands, including those in the south of Britain.

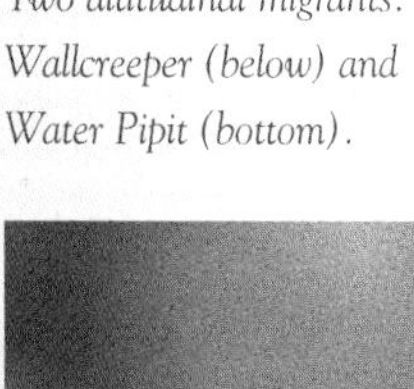

*Two altitudinal migrants: Wallcreeper (below) and Water Pipit (bottom).*

### Up to the Treetops

*Some people have suggested, with tongue in cheek perhaps, that the 'altitudinal migration' of the Capercaillie should be considered the shortest migration of any bird. In summer and autumn these large grouse feed mainly on the ground, feasting on shoots and berries, but in winter they switch their diet to conifers' needles, which they find high in the trees. A movement from ground level to the treetops may not seem like a migration, altitudinal or otherwise, but perhaps it is!*

## Partial Migration

When people refer to a bird as a 'migrant', the usual understanding is that every member of its species will undertake the journey prescribed for its kind. For example, all Cuckoos are expected to depart from Europe in the autumn and return in the spring, without any dissenters for either journey. This, if you like, is 'normal' migration. But a good many species do not follow such strictures, and instead exhibit what is termed **partial migration**. Within their populations, some individuals migrate and some do not. These individuals live in the same places in the spring and summer, and they cannot be separated by sight or by the way they behave during the breeding season, but as soon as autumn comes they show widely divergent migratory behaviour.

Most birdwatchers will appreciate the existence of partial migration without necessarily knowing its name. If you were asked, 'Is the Meadow Pipit a migrant?', you might recall seeing them flying overhead in October, obviously going somewhere, while at the same time realizing that they are classed as resident in Britain. This conundrum is explained by partial migration.

The phenomenon arises when there is a fine balance between the advantages and disadvantages of migrating. For example, a bird that remains on its breeding grounds will have a good chance of surviving if there is a mild winter, and by doing so it will reap the benefits of not migrating because it can acquire a territory earlier

*These two Meadow Pipits are in each other's company, but one might be a migrant and the other not. It's in the genes!*

*Experiments with Blackcaps has provided evidence that the tendency to migrate is genetic, leading to the phenomenon of partial migration. In cross-breeding experiments the tendency to migrate shown by a parent can be gradually bred out of the offspring within a few generations.*

next year than would be possible for a migratory individual. Moreover, it potentially could fit in more breeding attempts than its migrant rival. However, these advantages are lost in a harsh winter, when a sedentary bird has a greatly reduced chance of survival. Should it succumb as is likely, its vacant territory will be filled by the migrant bird that arrives after the winter, and this individual will now reap the benefits of its migratory behaviour. The vagaries of climate transfer advantages to one set of birds or the other over the course of time, which means that partial migration persists within the population as a whole.

A great number of species are partial migrants. A particular species may also be a partial migrant in one part of its range and a complete migrant in another. In fact, some birds may exhibit migratory behaviour at one climatic extreme of their range and resident behaviour at the other, with partial migratory behaviour in between.

Berthold (2001) memorably describes partial migration as 'a form of evolutionary turntable between sedentariness and migratoriness'. In other words, it allows a bird population to switch between one survival strategy and the other by means of natural selection. Birds that adopt the more successful strategy – whether remaining resident or migrating – will raise more offspring. So if the climate of an area changes over the years to become consistently milder or more severe, shifting the advantage of being a resident or migrant decisively one way or the other, then having the system of partial migration within a population enables it to adapt and survive over the long term.

# Differential Migration

If we take on board the fact that birds do not normally migrate in family groups (*see* pages 42–44), then it should come as no surprise to discover that different classes of bird – males, females and immatures – regularly undertake different journeys. Immature birds, for example, often migrate further than adults, and once they have dispersed from their parents' territories they might also leave the wider breeding area earlier or later than the parental generation (for more on the dispersal of immature birds, *see* page 23). Such a divergence of migratory behaviour within a species' population is known as **differential migration**.

Differential migration has given the Chaffinch its scientific name of *Fringilla coelebs*, the 'bachelor finch'. In Sweden, where the name was coined, only the females migrate in the autumn, leaving the males to sit out the winter in single-sex flocks. Many other species show similar patterns. In Britain, for example, most of the Smew that come to spend the winter are 'redheads' – birds in female or immature plumage. Only a few of the hardier, smartly dressed males deign to come so far south and west.

*Spot the female! The two sexes of Pochard (these are males) winter in different areas.*

Why should birds of different age and sex make different journeys? One possible reason is that they are avoiding competition. It might not be in the interests of the species as a whole for them to compete for the same food in the same places, and so a migratory system effectively keeps them apart. Another important reason, especially with regard to the polarized behaviour of males and females, is that a male benefits from arriving on the breeding grounds early to set up a territory, whereas a female is not lumbered with such considerations. Knowing instinctively that an

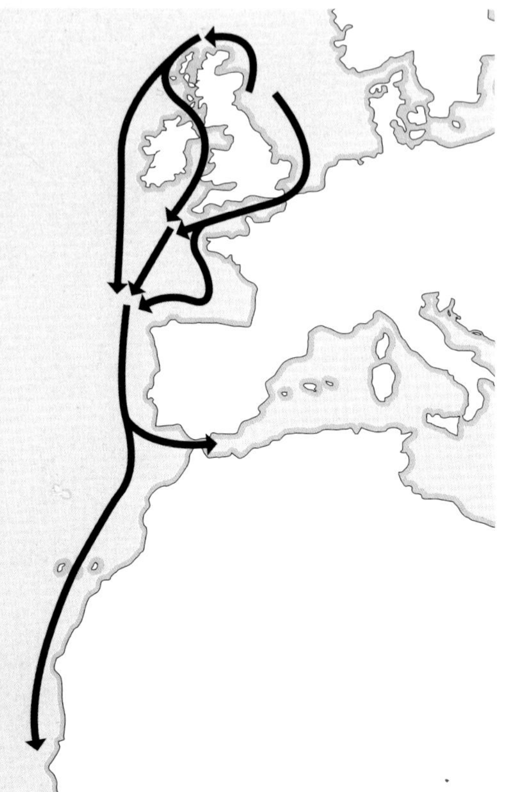

early arrival will enable them to establish a bigger and better territory, with dramatically enhanced breeding potential as a result, the males of many species winter as close to their breeding grounds as possible. The same attraction ensures that even among long-distance migrants travelling between continents, males usually arrive a week or two before females.

A special form of differential migration goes by the name of **retarded migration**, or **graded migration**. In some species, including seabirds such as the Northern Gannet and Sandwich Tern, young birds in their first season undertake their outward migration as normal, but either do not make the return journey at all in their first spring (retarded migration) or travel only part of the way back (graded migration). Only when they are a little older do such birds make the full return journey. In these cases, the wintering areas may act as a kind of nursery where youngsters can avoid the risks of travel or an uncertain climate until they are better prepared to do so.

*The migration of the Northern Gannet. Birds from all age classes may go south as far as West Africa, but birds in their first winter usually travel farther south than adults. The following spring all adults return to the breeding colonies, but the first years may remain in the wintering areas or travel part of the way back.*

*Lesser Black-backed Gulls show a degree of differential migration. First winter birds (right of three) tend to move further south than adults (top) or second winters (bottom).*

# Dispersal

Dispersal falls within that grey area of bird movement – is it migration, or isn't it? But however it is defined, there is no doubt that dispersal is a very important and almost universal phenomenon.

When young birds leave the care of their parents and become independent, they usually do not hang around in their natal territory but instead go off exploring, often for several months; it's the avian equivalent of a student's gap year! In most cases, they do not go very far, perhaps only a few kilometres. They typically join flocks of other juveniles and wander the local area, with no clear destination and no specific orientation. Their wanderings are spawned by an inherent urge and, at least at first, these are unforced by external factors such as lack of food or space. This movement is dispersal or – to be more specific – **post-juvenile dispersal**. It has also been called 'dismigration'.

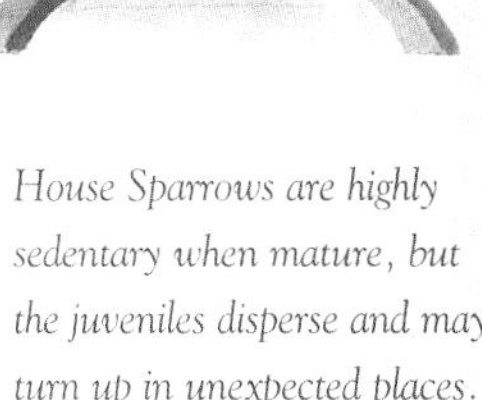

*House Sparrows are highly sedentary when mature, but the juveniles disperse and may turn up in unexpected places.*

The main purpose of post-juvenile dispersal is that it enables young birds to acquaint themselves with the neighbourhood with a view to a future breeding attempt. It is important to realize that post-juvenile dispersal occurs in both migratory and sedentary species of bird. In non-migratory species it has the additional benefit of promoting genetic exchange, ensuring that close relatives spread out from each other to avoid inbreeding. It might also enable normally sedentary species to expand their ranges into new areas and thus prevent overcrowding in traditional localities. While undoubtedly conferring these same benefits on migratory species, dispersal also allows the young birds to become so familiar with their local neighbourhoods that they then find it easier to return after their 'true' migration in the spring.

*Two juvenile Blue Tits soon to disperse.*

There are occasions when dispersal takes young birds much further than a few kilometres. In years of exceptional breeding success, immatures of highly sedentary species such as the Dunnock or House Sparrow may turn up at coastal ringing stations far from their parents' territory, in what presumably is an attempt to escape overcrowding. Some species disperse widely because potential feeding sites are separated by long distances: immature

Black-crowned Night Herons, for example, scatter in all directions looking for suitable ponds and wetlands. These movements may take the birds 1,000km or more. And that, let's face it, should certainly qualify for the term 'migration' in anybody's book.

*Juvenile birds often gather in mixed species flocks and roam the neighbourhood in a wandering movement that is known as post-juvenile dispersal*

## The March of the Little Egret

*The Little Egret has colonized Britain rapidly in the last ten years, going from no breeding pairs at all in 1995 to a maximum of 111 in 2002, spread over 18 sites. The initial colonization arose from the post-juvenile dispersal of young birds in 1989 after a successful breeding season across the English Channel.*

# Nomadic Movements

*Europe's nomad, the Common Crossbill.*

To and fro migrations are well known and easily understood. Much less obvious, but equally fascinating, are movements in which birds perform a 'to' but not a 'fro'; in other words, they leave a given area at a certain time, but then do not return. Their migration is one-way, and when it has finished, the birds settle down, breed, and remain in their new area for an indeterminate period of time. It is the equivalent, in human terms, of moving house.

Such movements are described as **nomadic**, and they are highly unpredictable. They are linked to the availability of food and may take birds in any direction. Nomadic birds do not have a fixed breeding range, but one that shifts location from year to year. In its simplest form, the nomadic lifestyle is a journey between areas of rich food supply. Human nomads wander from place to place looking for good grazing areas for their livestock, and nomads of the bird world do something similar.

There are very few truly nomadic birds in Europe, but one with good credentials is the Crossbill. This species has a highly specialized food supply: it is completely dependent on the seeds of conifers such as spruce, and eats very little else. Spruce is not a very dependable crop – some years see good seed production, but others see little – and unfortunately for the Crossbill, the trend may be synchronized over wide areas. So the Crossbill evacuates its breeding sites in poor years and travels in any direction, eventually settling down in a new place where the conifer seeds are ripening. The Crossbill's migration is a one-way ticket and although it may take place in any month, it is most often observed in July, after the birds have bred.

Nomadic behaviour is much more common in the arid regions of the world, where the availability of food is governed by erratic rainfall. On those rare occasions when the desert blooms, all kinds of nomadic birds will suddenly appear, apparently from nowhere, to take advantage of the temporarily abundant food supply and to breed. In Australia, for example, where rainfall is unpredictable throughout much of the arid centre, nearly a third of all breeding species are nomadic to some extent. They may follow the flowering of blossoms, as honeyeaters do, or the seeding of grasses, like finches and Budgerigars. They may also be more specialized. The Mistletoebird, as its name implies, exploits the various species of

*The arid centre of Australia, with its unpredictable rainfall, plays host to a large number of nomadic species. These are Zebra Finches.*

mistletoe found in Australia as each produces fruit, and different species do this at different times in different places.

Nomads have developed some unusual adaptations to cope with the erratic appearance of suitable breeding conditions. For example, both Crossbills and Australian Zebra Finches can breed when only a few months old and still in juvenile plumage. Whereas most birds are physiologically adapted to reproducing at a certain time each year – their sexual organs shrivel after the breeding season, only to grow again in time for the next – nomadic birds must be able to breed at the drop of a hat, so to speak. If they miss the opportunity at any time in their lives, they might never have another chance again.

## Irruptions

**Irruptions** are migrations that occur irregularly and take bird species away from their normal range. Birds 'erupt' from one place and 'irrupt' to another, at unpredictable intervals. These movements are similar to nomadic migrations, except that they are not directed towards new breeding areas but away from untenable ecological conditions, and birds that are involved in irruptions sometimes return to their usual breeding sites after a few months, unlike nomads. Furthermore, irruptions are less frequent than nomadic movements, often occurring after many years of sedentary behaviour or normal to and fro migration.

*The Waxwing erupts from its northern breeding grounds when its population is high and food supplies are low.*

As far as birdwatchers are concerned, irruptions are exciting events. Species that are very rare in a particular region suddenly turn up unannounced, often in large numbers. In Europe most irruptive species are from the far north, including such characterful birds as Waxwings, Pine Grosbeaks, Nutcrackers, Bramblings and Snowy Owls. A few species also come from the eastern steppes, such as Rose-coloured Starlings and Pallas's Sandgrouse. In addition, alongside these headline-grabbing superstars there are a few more familiar faces, among them Great Spotted Woodpeckers, Jays and various species of tits. But whatever the species involved, birdwatchers love the unpredictability of irruptions, and enjoy seeing flocks of birds that are out of the ordinary.

The trigger for an irruption is not always clear, but studies on Waxwings have shed some light on the phenomenon. It seems that Waxwings erupt when their population density is too high and food availability at the end of the breeding season is too low. Such a coincidence of events happens only occasionally, because in some years large populations will be adequately sustained by plentiful food in the normal range, and in other years a poor food crop may suffice for a relatively small population. The extent of an irruption and the number of individuals involved presumably depends on the scale of

*Very occasionally large numbers of the Northern, white-headed race of Long-tailed Tit erupt south or west. A few turned up in Britain in the winter of 2003/2004.*

*The Siberian race of the boldly-coloured Nutcracker irrupts into Europe at highly irregular intervals, and some individuals involved in the movement may remain to breed far outside their normal range. The small population in Belgium colonised this way.*

the breeding boom and subsequent food shortage, with the best breeding seasons and most acute food deficits leading to the most spectacular evacuations. Whatever the cause, Waxwings are such colourful and distinctive birds that their irruptions are hard to miss; the first one was recorded as long ago as 1413.

The food source that matters most to Waxwings is rowan berries. These appear in August and September, when the Waxwings have just finished breeding and might still be in family groups. If there are not enough berries to go round, many families are forced out, and this situation may recur later in the autumn and winter when supplies are further affected by heavy falls of snow. Gradually, Waxwings appear further and further away from their normal wintering range. They begin to subsist on other types of berry and, in Britain at least, seem to be forever drawn to superstore car parks, the latter styled with their modish plantings of berry-bearing bushes.

Certain birds of prey also display irruptive behaviour, although the pattern of their movements can be much more cyclical than that of berry- and seed-eating birds. This is because, for reasons that are poorly understood, several species of small rodent exhibit sharp fluctuations in their numbers from year to year. For example, voles and lemmings become abundant every four years in the north of Europe, and then their population crashes before building up again. Not surprisingly, such a crash triggers an evacuation of the predators that depend on them, carrying species such as the Short-eared Owl and Rough-legged Buzzard to unfamiliar places well south of their usual wintering range.

# Moult Migration

Most species of bird time their migrations around the changing of their feathers. They might, for example, moult before they set off, as most small birds do; or they might wait until they have reached their wintering grounds, like some waders, gulls and terns. They might even, like another group of waders, moult during a pause in their journey at some favourable staging area.

However, a few birds perform a quite different journey – a migration to a particular area with the express purpose of moulting upon arrival. Such journeys are called **moult migrations** and are seen in various wildfowl, grebes and seabirds. Moult migrations are typical of birds that swap all their flight feathers roughly simultaneously, leading to a brief period of flightlessness during which they are very vulnerable to predators. Ideal sites for moulting, which must provide exceptionally safe conditions and plenty of readily available food, are at a premium and are thinly scattered, hence the need to make a special journey.

*The Canada Goose (top) has developed a moult migration since being introduced here from North America in the eighteenth century. Large numbers of birds from the midlands northward migrate to the Beauly Firth in Scotland after breeding and moult there. Interestingly the closely related Barnacle Goose (bottom) lacks any moult migration at all.*

*Breeding adult Mute Swans with young stay put and moult within their territories, but non-breeders migrate to a small number of traditional moulting areas.*

Interestingly, in Europe a lot of moult migrations take birds northward in autumn, the reverse of what one might expect. But perhaps this is not so surprising, after all. In the north there are probably more lakes and seas suitable for moulting, and more abundant food of a particular kind – such as the fresh plant growth that geese require. There are also more daylight hours in which to feed.

*The Shelduck's summer festival – moulting on the Waddensee.*

## The Moult Migration of the Shelduck

*After breeding, adult Shelducks from throughout north-west Europe, including Britain, leave their breeding grounds and converge on a small area off the coast of northern Germany known as the Waddensee. Up to 100,000 Shelducks may be found in the food-rich shallow waters and mudflats of this area. They come from all directions, including north from Scandinavia and south-west from France.*

*The Shelducks leave for the moult on July and August evenings at around sunset, and make the flight in one go. They complete their moult in a few weeks, but then, apparently relishing the easy life, they become distinctly reluctant to leave the area. A few diehards are still present in October.*

*The moult migration is the Shelduck's only annual movement: the species does not have special wintering grounds. Instead, the British population makes a very slow homeward movement, with some birds not arriving back on their breeding grounds until December or even January.*

# All-year Bird Movements

*Redwings come to Britain in October and return north in March and April.*

By now you have probably realized that, far from being confined to the so-called 'migration season', bird movements actually occur throughout the year. To demonstrate this point further, let's look at a year in the life of Britain's birds.

The earliest spring migrants tend to arrive in March, and from then on the migration is very heavy until May. Certain individuals and species, such as Marsh Warbler, are still settling in during June, and at the end of that same month a few failed breeders from the Arctic – especially waders – begin their return migration. July is a time when many fledglings disperse away from their breeding areas and some species of wildfowl begin their moult migration; Crossbills undergo their main movement, and more non-breeding waders swell the numbers already in the UK.

*Sand Martins – amongst the earliest summer visitors to arrive in Britain, in early March.*

The great autumn movement southward begins in August, and birds such as Swifts, Nightingales, Pied Flycatchers and adult Cuckoos have all gone by the month's end. During September and October there are enormous movements of birds, both away to the south and into the country from the north and east, and some of these comings and goings continue into November. During the rest of the winter – in December, January and February – many birds are forced into Britain by bad weather on the continent; if adverse conditions persist and spread to this country, there will be escape movements from these islands, too. Finally, a small number of summer visitors to the UK actually make landfall before the end of February.

*Marsh Warbler – perhaps the last of our summer visitors to arrive, often in June.*

# 2 THE MIGRATORY JOURNEY

*The purpose of a journey is one thing; what about its specifications? If we were about to embark on a trip, we would want to know all about it in advance: how far we had to go, how fast we would be going, when we would leave and arrive, would food be served, would anyone we know be coming along? We don't know whether birds ask these questions, or if they have the answers; but we can ask them on the birds' behalf, and that's the purpose of this chapter.*

*House Martin and Swallow in the autumn.*

*Some Ruffs travel 30,000km a year.*

## Distance

Bird movements span the globe. Many species familiar in Britain are transcontinental travellers that commute thousands of kilometres between their breeding and wintering areas. Indeed, we could properly think of many of them as African birds that venture north for a short breeding season. About 200 species migrate between Europe and Africa, and more than 300 move between North and South America. Birds also travel between northern and southern Asia, between Australia and New Zealand, and across the Equator. Distance – even for quite small birds – seems to be no object.

Waders, too, are great travellers and many clock up long journeys, including the population of Ruff that moves to Africa from Eastern Siberia, covering 30,000km on the round trip each year. Among birds of prey, Swainson's Hawk may travel as much as 12,000km each way between Alaska and northern Argentina. And even a bird as tiny as the 9-cm (3.5-in) Ruby-throated Hummingbird, which weighs a mere 3g (⅛oz), can fly as far as 6,000km from its breeding sites in southern Canada to spend the winter in Central America.

### The Longest Journey

*Which birds undertake the longest migratory journeys? The record holder is the Arctic Tern (see page 112) that migrates from the Arctic to the Antarctic and back, regularly covering 35,000km in a single year. Several species are not far behind: many shearwaters fly nearly 30,000km in a year, and even some passerines, such as the Northern Wheatear, do much the same overall mileage – especially those that migrate from Africa to Greenland or Alaska and back.*

*Small bird, big migration: the Ruby-throated Hummingbird can cross the Gulf of Mexico in a single flight.*

Impressive these feats might be, but some may be achieved at quite a leisurely pace, with frequent stopovers, often of several days' or even weeks' duration. The flights that really catch the imagination are the non-stop transoceanic marathons. Once again, waders dominate the world records, almost in the manner of former East German athletes! For example, Pacific Golden Plovers leave their Alaskan breeding quarters in August each year and do not touch down again until they reach the Hawaiian Islands – 4,500km and 100 hours' flying time away.

*Pinpoint navigation – some Alaskan Turnstones migrate 5500km non-stop to mid-Pacific islands, a journey that allows for negligible margin for error.*

Alaskan Turnstones are just as extraordinary, travelling between the Pribilof Islands to the west of Alaska and the Midway group to the west of the Hawaiian chain. It is thought that some individuals actually ignore the Midway Islands and instead continue to the Marshall Islands in the south Pacific, a non-stop leg of some 5,500km in all. The migration of the rare Bristle-thighed Curlew is equally astonishing: from their breeding areas in Alaska, some birds are thought to overfly the Hawaiian Islands and go straight to islands in the south-west Pacific more than 6,000km away.

These are flights of awe-inspiring proportions. Imagine flying for 100 hours without stopping, at a height of perhaps 6,000m, aiming for a relative pinprick in the midst of a mighty ocean. What these birds must see and what they must experience along the way! It is only thanks to our comparatively new invention of the aeroplane that humans can begin to appreciate what these birds have been doing every year for thousands of years.

*Young Guillemots may swim 700km away from their colony after they have left the nest, often accompanied by their male parent. One chick is known to have covered 356km in 15 days.*

## STAGING AREAS

*Pied Flycatchers from throughout Europe begin their migration by flying to northern Spain and Portugal.*

We might be awestruck by the non-stop flights of the greatest bird migrants, but relatively few species are involved in these record-breaking feats. The vast majority of migratory birds perform their migration in stages, flying from one place to another and then resting and feeding before moving on again. It is the same for us when we undertake long car journeys: we stop periodically at motorway service stations to refuel ourselves and our vehicles.

The first stages of the southward migration of a small bird journeying from Europe to Africa are distinctly modest. The initial few legs frequently amount to only 50km or so each, which requires only a couple of hours of flying time after dusk. Gradually the distances flown increase, until the bird, having reached the major barrier of the Sahara Desert, may fly all night or all day without a break. But the fact is that, particularly on the outward journey, bird migration is often a stop-start affair, with plenty of time being spent between the point of departure and the final destination.

This explains the importance of **staging areas**, which are effectively service stations for birds. Over many centuries of migration the routes plied by different species acquire their own metaphorical 'wheel-ruts' and 'wayside inns', and it becomes second nature for members of a bird species to stop at particularly amenable localities. Pied Flycatchers, for example, whether they are breeders from Russia or Wales, invariably make a major stop in northern Spain and Portugal to build up their fat reserves before setting out on the next part of their journey south.

*Thanks to good supplies of food available en route, Brent Geese arrive on their breeding grounds well fed.*

Staging areas are just as important, if not more so, for birds on their return migration. Brent Geese, for example, which breed on large islands in the Arctic Ocean, only have access to their breeding sites for a hundred days each year due to snowfalls either side of this ice-free interval. When the geese arrive, it is quite common for there to be little or no grazing on-site because the tundra vegetation has yet to sprout after the thaw. So the geese make sure that they reach their destination with plentiful fat reserves that are enough to sustain them until food becomes available. And where do they build up those fat reserves? The answer is on their staging areas, notably in the Waddensee, off Germany.

Staging areas are special places that obviously make an important contribution to a migrant's overall survival prospects. However, they can pose a real problem for conservationists. A rare migrant may need to spend only a few days or weeks at a particular place each year and it can be quite difficult to convince the relevant authorities that an area is worth preserving for such a brief patronage.

## SPEED

Migration is a marathon and not a sprint, to adapt the time-honoured phrase of sports journalists. Unless a bird is on a long flight over the ocean, or it is escaping some exceptionally severe weather, speed of flight is not especially important. The thing that matters is arriving safely.

This is particularly true on the outward migration, which can be a distinctly leisurely affair. Many common summer migrants, such as the Swallow, take at least three months to travel from northern Europe to Africa; Marsh Warblers take even longer – sometimes as much as six months. Such birds are not in a hurry. Only during the spring migration do they behave as if they are in a race; males may be forced to reduce the time they take to get to their destination in order to arrive early and claim the best and largest territories.

But what of a migrant's speed during each stage of its journey? Birds, predictably enough, do not try to fly as fast as they can, but they do try to fly as efficiently as they can. More specifically, they go at a speed that allows for the minimum energy consumption per distance covered. It would take some complex sums for us to calculate what speed this should be, but birds manage to bear out the theory without recourse to a computer! Perhaps it is simply a speed that 'feels right' for them.

The speed at which migrants fly varies between species. Smaller birds fly at an average speed of 30–40km/h; thrushes travel somewhat faster, at about 45km/h; pigeons travel at 60km/h; and some waders and ducks can fly at 70–80km/h. The Swift, despite its name, is something of a migratory slouch, managing only about 40km/h.

These speeds are all ground speeds and take no account of the wind. One would expect that a bird would reduce its speed when assisted by a tailwind in order to save energy, and this is the case. Conversely, when a migrant is flying against a headwind, there is some evidence to suggest that birds fly faster. However, in general this increase in speed is difficult to measure, at least in part because birds prefer not to migrate into the wind if they can avoid it.

*Despite their name, Swifts don't fly particularly fast when migrating.*

One final point to make about migration speeds is that because of the reduced density of the air at higher altitudes, a high-flying bird will increase its energy efficiency if it increases its speed. Many observations bear this out. For example, birds at altitude have been confirmed by radar studies to fly faster than individuals of the same species at ground level.

**WILD WESTERN**

*A Western Sandpiper carrying a small radio transmitter once travelled a distance of 3,000km from San Francisco to Alaska, in less than 42 hours, at an average speed exceeding 64km/h.*

*Flock of Western Sandpipers showing the typical long, pointed wings of the long-distance migrant.*

## HEIGHT

While on the subject of altitude, how high do birds normally fly when they are migrating? The answer differs from place to place, and also between night and day. Interestingly, day-flying migrants tend to fly lower on average than night-flying migrants, for reasons that are as yet unclear.

In the 1970s, an enterprising ornithologist called Frank C. Belrose carried out what must have been an exhilarating series of experiments. He took to the air in a light aircraft by night and counted the birds as they whizzed past, illuminated by a series of searchlights. By flying at various different heights over the same beat, night after night, he managed to obtain an idea of the density of birds flying at certain altitudes. His findings were highly informative. The greatest density of birds was to be found at 300m above ground, with plenty at the 150m level and at the 450m level. Very few were found above 900m and, of particular interest, not a single bird was recorded below 30m.

Radar studies have added more weight to these observations: it is now thought that about 50 percent of small migrants travel at less than 400m above the ground. Occasionally there are exceptions, with even some small species attaining mind-boggling heights of

6,000m. But if you look up into the night sky on an autumn night, you can safely assume that most of the birds are just a few hundred metres above you.

*An audacious way to count night-flying migrants – Frank Belrose's spotter plane.*

Interestingly, radar studies have also revealed that birds' flying altitude varies during a night of migration. Night travellers usually leave just after dusk and quickly gain altitude, until they reach a maximum in the early part of the night and maintain it until about midnight. The altitude then drops gradually until the birds land at dawn.

## Hitting the Heights

Of course, there are occasions when a migrating bird needs to fly much higher than usual. The most obvious example is during a flight over high mountains, and no bird is more famous for its feats of altitude flying than the Bar-headed Goose. This species breeds on mountain lakes in Central Asia and winters in the lowlands of northern India, a journey that inevitably takes it directly over the Himalayas. It is a relatively short flight, but the Bar-headed Goose does it in style, passing serenely over the

*Bar-headed Geese overflying the Himalayas.*

highest peaks; it has been seen by mountaineers flying over the summit of Mount Everest at 8,849m and in laboratory conditions it has been able to tolerate simulated altitudes of 10,000m.

Birds also fly to great altitude during long flights over the sea. For example, a good many species – large and small – brave the long 4,000km stretch of ocean between the north-eastern United States and the West Indies. It is the quickest way there, following the so-called 'great circle' route rather than the much longer land route. During the stage over Puerto Rico the average height for the travellers can be as high as 5,000m, although it ranges between 1,100m and 6,800m. Similar flights have been noted for birds undertaking other ocean crossings, and also deserts.

There is one important factor still to consider: wind. In 1967, a group of Whooper Swans was discovered by radar flying over the eastern Atlantic between Iceland and Northern Ireland at an altitude of 8,200m. Their movement was in midwinter, probably an evacuation from difficult conditions, and it coincided with a strong blast of cold northerly winds and snow showers. Measurements taken at the time indicated that at the birds' height, they were safely above any snowfalls and in the midst of a powerful tailwind that touched 180km/h. No doubt the swans were using this wind to propel them rapidly to a milder destination.

In fact, the wind is probably a major factor in determining how high birds fly during normal migration. That would hardly be surprising, since the strength and direction of the wind is the primary atmospheric consideration for flying birds.

### Flying High

*Why do birds fly so high on occasion? The 'why' is quite difficult to answer – perhaps birds actively seek the cooler temperatures at these heights to avoid overheating during the exertion of flight. Or perhaps it is more efficient in terms of energy consumption to pass through the thinner air. Possibly it is safer up there. The answer remains elusive.*

# Day or Night?

Migrant birds embraced the 24-hour-a-day lifestyle long before we did and during the main migration periods the skies see non-stop bird traffic both by day and by night. However, many species have a strong and specific preference towards the former or latter time.

The birds that most clearly need to migrate diurnally (by day) are species such as storks and large birds of prey that rely on soaring in **thermals**. Thermals are bubbles, or updraughts, of rising hot air that develop in the heat of the day – but not at night. Swallows, martins and swifts also migrate in daylight, perhaps because this enables them to feed in transit. Many short-distance migrants do so, too, including larks, pipits, wagtails and seedeaters such as finches and buntings, although the reasons why these groups should move diurnally are not so obvious.

*House Martins and Swallows usually migrate by day.*

The majority of migrant birds, however, including just about every insectivorous long-distance traveller, are in the nocturnal camp. They include warblers, flycatchers and thrushes among their number, as well as larger birds such as waders and wildfowl. In the normal course of events, not one of these species is nocturnal and it is only the act of migration that switches them over. From this we can conclude that migrating at night must have some special advantages. But what might these be? In common with much about bird migration, there are plenty of theories, but no clear-cut answers.

The most compelling arguments are to do with physiology. It is cooler at night, so migrating at this time could protect birds from overheating and dehydration, which must be a possibility in view of the amount of energy they are continually expending. Flying at night is also atmospherically favourable: the air is cooler and denser at night, and it is usually less windy, all of which can help migrants to save energy.

Of course, migration at night also leaves the day free for other activities, notably feeding and feather care. Perhaps by compartmentalizing its time – travelling at night and refuelling by day – a bird can cover the ground more quickly, especially if it is traversing difficult terrain. It sounds very efficient, but the question is, does a bird *need* to do it? Studies tend to show migration to be a restrained and intermittent process, with stops for rest

Ready to go!

and refuelling. It seems unlikely that the necessity to move on quickly should have led to the widespread adoption of nocturnal migration.

Another common theory is that flying at night keeps migrants safe from predators. Perhaps it does, but it is doubtful that this could be the whole explanation for night migration, for it would have to be an escape from slaughter on a huge scale, and this simply has not been recorded among even daytime migrants. Furthermore, the night may hold its own dangers, such as owls and possibly even certain species of bat.

Some birds use the stars to help them find their way, so maybe this could be the reason for migrating at night? This seems especially unlikely because if you don't go out at night you can't see the stars – the birds must have been out at night first to notice that the stars were there and that they moved around the night sky. A more plausible explanation for the so-called **star compass** (*see* page 65) is that it arose as an orientation tool as a result of birds migrating at night, not the other way round.

*By flying in so-called 'V-formation' birds can save energy. When a bird flaps, its wings quite naturally push air down, but this creates a corresponding updraft just off the wing-tip. If another bird can follow along at the right place, therefore, it can tap into the updraft and obtain lift for free; hence the formation. Incidentally, one bird doesn't do all the hard work; the 'leaders' change places regularly.*

## Keeping Good Company

From a human perspective, the hardship of a long journey is considerably eased by a bit of company, and it is natural for us to assume that birds usually migrate with family or colleagues. The benefits are obvious: more pairs of eyes could, for example, help with finding food, or with locating suitable habitat along the way; the individuals could also pool their navigational skills to make sure that they are going the right way. The youngsters could 'learn the ropes' from older birds that have travelled before. And being in a flock also has some energy-saving benefits if the group adopts certain flying formations (*see* panel).

Despite all this, most birds shun such luxuries and travel on their own, at least in the sense that they do not have a constant companion throughout their journey. The most obvious example of this is provided by the Cuckoo. Young Cuckoos are, of course, not brought up by their parents; the adults migrate from the breeding areas long before the juveniles, so they cannot possibly guide their progeny to the right wintering site. An immature Cuckoo's foster parents are not any help, either, because they may be non-migratory, such as Robins or Dunnocks, or they may be migrants that have completely different wintering areas. A young Cuckoo will not see any siblings and it will not normally migrate in flocks. In short, a young Cuckoo makes its entire journey unaided.

*The juvenile Cuckoo must find its way to its winter quarters in Africa entirely alone.*

Surprisingly, the Cuckoo's extreme isolation in terms of migration is not unusual. Most of our small birds are wholly responsible for their own migration, relying entirely on their inner drives and physical abilities to get them to their destination. Swallows, Willow Warblers, Spotted Flycatchers and Blackcaps, for example, are all go-it-alone migrants.

That's not to say that all or many of these species avoid members of the same species in transit – they don't. A lot of birds do travel in informal flocks, both by day and by night. At night members of these flocks call to keep in contact, and by day they can see each other. And being in a flock does confer all the benefits just listed. The point is that birds do not stick with 'friends' and do not make allowances for other birds. The flocks they form are short-lived and incidental. They might even be coincidental; after all, good weather conditions for migration are at a premium, and draw birds to the skies like shoppers to the January sales.

*Siskins may pair up in winter flocks and undertake their spring migration together.*

## In It Together

However, there are some exceptions to this rule. A number of larger species, including geese, swans and cranes, travel in families, at least on their outward migration. Adult terns may spend part of their journey with their offspring, and might even take time out to feed them occasionally in transit. The male Guillemot disperses out to sea with its chick once the latter has jumped from the nest on a cliff ledge, and it may swim tens or hundreds of kilometres alongside. Finally, a good many species, particularly those with short journeys, undertake their return migration in newly formed pairs, having established their pair-bonds on the wintering grounds. A lot of ducks do this, and the behaviour has also been proved in several passerines, such as the Siskin.

When birds meet up and form pairs on their wintering grounds, the new partners will often have quite different origins. A Goldeneye from Scotland, for example, might take a fancy to a bird from Sweden. Once the bond has been formed and return migration beckons, it is now a case of 'your place or mine' and a 'decision' must be taken. Normally, the males follow the females back to their natal area, resulting in the drake taking a different return journey from the one he would have taken had attraction not intervened. There is a name for this special amended journey: **abmigration**.

### Two by Two

*The Bearded Tit (above) is one of the few small birds to migrate in pairs, or groups of pairs. This is to allow for an even sex ratio during the colonization of new breeding sites.*

## Risks

Migration is by no means as dangerous as is popularly perceived. Remember that migratory journeys are primarily for survival. Nonetheless, there are risks attached, and they come in two guises: there are the dangers inherent in the act of travelling, including physical stress, the potential to get disoriented, and the vagaries of the weather; and there are the risks that come with the locations visited along the way. These may include predation, being shot, or starvation through lack of food or water.

In order to combat physical stress, birds take great care to prepare for their journeys. Ideally they leave only when they are ready, although they may be held up by poor conditions for migration, or they may hasten away due to a sudden drop in the amount of food available on-site. They also try to reduce physical stress where possible. Many small birds start their migration in short trips of 50–100km and only increase their flying times gradually. They may stop off for some days before moving on.

*Turtle Doves are frequent victims of hunters during their migration.*

Any migratory journey, especially over water, is particularly prone to weather problems. If birds encounter persistent side winds they may drift off course and end up in the oblivion of the ocean. Birds going in the right direction may encounter strong headwinds and succumb to exhaustion before they make landfall. Heavy cloud, fog and rain can all disorient birds. So can bright lights. Migrant birds used to die in their hundreds when attracted to old-style beam lighthouses, flying towards them like moths to a flame. They still occasionally get burned in gas-flares, and hit tall and well-lit buildings.

Certain places cause particular problems for migrant birds. Shamefully, many migratory birds are still shot by people in Europe, especially in France, Italy, Cyprus and Malta, and this can also be a problem further along the migration route, in Africa. The Sahara Desert presents a different kind of barrier. It was once thought that migrants flew over its barren wastes in one go, but now it appears that they cannot easily fly over it by day because of the risk of overheating. So they rest by day where they can – under rocks or small bushes, for example – and fly by night.

Wherever a bird travels it is potentially at risk from predators – especially other birds. Every site has its local predators and one or two species, including Eleonora's Falcon (*see* page 120), specialize in catching migrants. Recently, it has been suggested that certain large bats may also take their toll on the hard-pressed travellers.

*Bright lights can be a hazard for migrating birds, attracting them like moths to a flame.*

*A Redstart rests in the scant desert shade by day.*

# 3 PREPARATION AND INTERNAL CONTROL

*Nobody would contemplate running a marathon without training first. In much the same way, birds do not wake up one day and decide to migrate to Africa. The journey is a long one, more than a marathon, and it cannot be faced without intensive preparation. It is therefore a good thing that, long before departure, birds' internal 'switches' register the future need to migrate and begin to get the birds into a state of readiness that will enable them to reap the rich dividends of a successful migration.*

*The Knot may double its body mass prior to migration.*

# Putting Down Fat

Many people would be horrified at the thought of what birds do to their bodies prior to migration. Not only do birds eat far more than normal, they also put on a great amount of fat as a result. Some small birds actually double their body mass – hardly the ideal basis for intense physical exertion, one might have thought. However, birds are physiologically quite different to us, and have a much higher metabolic rate. The fat does not remain for long because it is the main fuel that birds burn up on migration. A migrant bird with lots of fat reserves is the equivalent of a jumbo jet with full tanks: it might be so heavy that it has trouble taking off, but once airborne the fuel is used up very quickly.

*Normally highly insectivorous, Blackcaps switch to a diet of fruit prior to migration.*

Fat is very much a migratory fuel. At most times of their life, birds mainly break down carbohydrates for energy, much as we usually do. Carbohydrate can be metabolized much more quickly than fat, and that suits a bird's fast-moving lifestyle better. But where fat really comes into its own is as a fuel for storage. For one thing, fat yields twice as much energy as the equivalent amount of carbohydrate. Secondly, it can be stored dry, without the need for water, and that saves extra weight. There are other metabolic perks, too: when fat is broken down it produces fewer waste products than carbohydrate, and when muscles run on fat they tire at a slower rate. So it is small wonder that birds take this elixir for their marathon journeys. Say it aloud – fat is good!

You might take a look at a migrating bird and say, 'It doesn't look fat to me'. And of course, it does not. The bird's feathers cover the extra bumps, rather like the maternity clothes of a pregnant woman. The fat is not stored in a single place, either, but distributed around the body, mainly under the skin, which reduces any potential aerodynamic imbalance. The difference would only become apparent if, like people who ring birds, you caught the bird and gently blew on the feathers on its belly to separate them. Then you might see the 'shirt of fat' that migratory birds wear. You'd also notice the difference if you weighed the bird; as mentioned above, many birds double their body mass at the height of their migration.

The process of putting on the fat and using it as fuel is not as straightforward as it may seem. It requires fundamental behavioural changes and also a variety of internal, chemical adjustments. For example, a bird preparing for migration has to eat much more than it did before, and it might have to change the type of food that it consumes. The bird's body must be able to cope with these things and has to effect the switch from carbohydrate breakdown to fat breakdown.

Anyone who has watched birds on migration will have a feel for these changes taking place. For one thing, the travellers appear to feed non-stop, dashing from bush to bush for insects, or gorging

*A Blackbird swallowing rowan berries – little and often is the key.*

themselves on berries long past the point at which you would expect them to be sick. This type of continual eating behaviour goes by the grand name of **hyperphagia** – 'Darling, I'm not eating too much, I'm exhibiting hyperphagia'. Strictly speaking, it alludes to the act of eating more often, rather than taking more at each meal. If conditions are right, a few weeks of hyperphagia can be enough to trigger the first migratory flight.

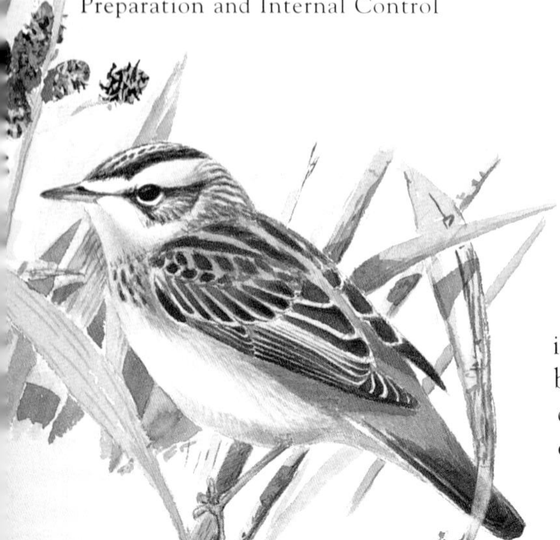

Many birds, as already implied, also change their diet prior to migration. This is especially obvious for species that are primarily insectivorous in the breeding season, such as thrushes and warblers. Suddenly their attention turns to berries and other fruits – frugivory comes into fashion! There is good sense in this. Berries are nutritious, many containing both carbohydrate and fat, and they are easy to digest. Even more important, they are highly conspicuous, accessible and often present in large quantities, so birds do not have to forage extensively for them. The relevant bushes 'want' their berries to be eaten in order to spread their seeds, and so offer them to migrants in abundance in the same irresistible way that supermarkets lay open their shelves to be plundered by card-carrying shoppers. Birds and bushes revel in the arrangement.

*Fuelled by aphids, Sedge Warblers fly to Africa non-stop.*

*Without aphids as their secret ingredient, Reed Warblers migrate in stages.*

## Insect Power

This does not mean that migrants forsake other types of food. Experiments have demonstrated, perhaps surprisingly, that on its own a diet of berries is not normally enough, especially in northern Europe where the fruits apparently do not share the nutritional pizzazz of their Mediterranean counterparts. Many insectivorous migrants continue to eat some insects as well. Presumably these contribute the extra necessary proteins that fruits lack, and insects also contain some fat of their own.

Certain kinds of insect evidently confer some intriguing fringe benefits. Aphids, for example, apparently greatly enhance the process of generating fat reserves and many warblers, in particular, devour them in large quantities prior to migration. For the Sedge Warbler, aphids seem to have the same effect as spinach does to Popeye! Having consumed enormous amounts of these tiny invertebrates, Sedge Warblers are propelled on their migratory flight all the way from Europe to West Africa non-stop, a distance of 4,000km that entails three days and nights of travel. Reed Warblers, on the other hand, which have a broader diet than

Sedge Warblers, receive no such fuel-injection, and their stage-by-stage migration is sluggish by comparison.

The metabolic changes that take place inside the migrant's body are highly complex. The switch from carbohydrate breakdown to fat breakdown requires not just the increased ingestion of fat and its deposition in the appropriate places, but also the conversion of a migrant's carbohydrate reserves, and subsequent intake of carbohydrate, to fat. Suddenly, fat is in demand everywhere throughout the bird's body, and the liver works flat out to produce it in a process known as **hyperlipogenesis**. At the same time, the body must ensure that all the necessary tissues cope with this new metabolic regime, not least the all-important pectoral muscles – the 'engines' that power the bird's wings. In some species, these muscles actually increase in size.

Every one of these changes is under the control of the body's army of enzymes and hormones, and the changeover of metabolism could be described as like a military coup. There is a new regime, pervading everything. The body experiences a flurry of biochemistry. If you looked hard enough, you could almost imagine smoke rising from the hard-pressed bird's overworked engines…

*A bird being weighed at a ringing station. Small birds often double their body weight prior to migration.*

## Migration Restlessness

Hyperphagia is only one of the external symptoms that you might notice in a migrant or migrant-to-be. Another symptom is 'migration restlessness', which is also known by the German word **zugunruhe**. It is most clearly demonstrated by captive birds during migration periods. At these times the captives seem unable to prevent themselves from jumping about their cages from perch to perch and fluttering their wings rapidly every few minutes or seconds. They give every impression of wanting to set off somewhere, like restless commuters fussing about as they wait impatiently for a late train.

This curious behaviour occurs during the night in birds that typically migrate nocturnally, and quite obviously it corresponds to the urge that drives wild birds to begin their flight and which keeps them going. One cannot do much about *zugunruhe*. You can alter a bird's captive environment as much as you like, but you cannot prevent the restlessness, which will carry on until the species' migration period is over.

*Zugunruhe* is a fascinating phenomenon to observe in its own right, but only by measuring it can its significance be appreciated. This can be done by fixing switches to perches so that when a bird lands on one it will automatically register to a counter. Over the course of the migration season a bird's continual shifting of position can be monitored. Then you will discover something amazing: the extent of restless behaviour corresponds to the length of a bird's migration. If you measure the exact amount of autumnal *zugunruhe* for different species and multiply it by the projected average speed of that species, the distance corresponds closely with a point in the middle of the natural wintering range.

So *zugunruhe* is a little like the energy stored in a clockwork toy, such as a car. You wind up the clockwork mechanism so much, and it takes the car so far; wind it up further and the greater quantity of stored energy makes the car go further than it did the first time. In the same way, the overall amount of migratory restlessness in a bird's pre-set programme energizes the bird to travel a certain distance and no further.

The crucial point about *zugunruhe* is that it is independent of external factors. It does not depend on the weather or on the bird's present circumstances, and it is not controlled by the bird itself. Whatever produces *zugunruhe* is internal. The chemistry of the bird switches it on, varies its intensity where necessary, and then switches it off again. It is an endogenous pre-set programme, and experiments have proven that it is genetically inherited.

*Zugunruhe* is related to a bird's other annual cycles, such those that induce moult or control breeding behaviour. These cycles are known collectively as the **circannual rhythm** and are probably calibrated by the changing ratio of day to night through the seasons. Yet at its heart 'migratory energy' is simply to do with changes in the bird's chemistry.

These findings are extremely important because they solve several migratory puzzles in one. If a bird's migration is pre-programmed, as *zugunruhe* suggests, this potentially could explain how a bird knows when it has arrived at its destination – it will suddenly run out of migratory energy. An internal programme could also 'set' the correct departure date, and the date for return migration, too.

## Setting Off

If you were about to undertake a walk from Land's End to John O'Groats, you would probably invest in a new pair of boots before leaving. The majority of birds do something very similar: they moult, that is, they exchange their old, worn-out feathers for new ones. This is necessary both for adult birds, whose breeding efforts will have taken their toll on the birds' plumage, and for youngsters,

*Swallows gather excitedly on wires before setting off on their migratory flight.*

whose juvenile plumage is only temporary and of somewhat inferior quality. So just about every migrant, of every species, sets off on its migration wearing new 'clothes'. Naturally, birds do not 'know' when to moult any more than middle-aged men 'know' when they will go bald. It is another internal programme, part of the birds' circannual rhythm mentioned in the previous section.

Let's review how all these programmes come together to co-ordinate a migrant's departure. The bird moults first, a process that can take up to a month. Towards the end of that programme, an internal switch induces hyperphagia, causing the bird to eat more often. A behavioural switch makes the bird look for high-energy foods such as fruit rather than just invertebrates, and much of this food is now converted into fat, which is stored under the skin. The bird's metabolism makes a complicated changeover from carbohydrate breakdown to fat breakdown. And then another trigger urges the bird towards a state of perpetual restlessness, the aforementioned *zugunruhe*, which will eventually send it off on the first leg of its journey.

The precise time of the bird's departure – and of every subsequent departure if the migration is completed in stages – is fine-tuned by two things. These are, firstly, the efficiency of the preparations listed above and the physical readiness of the bird and, secondly, the arrival of suitable weather conditions for

*Fuelled with berries and super-fit, these Lesser Whitethroats await the right weather conditions to trigger their final departure.*

migration (*see* chapter 5). In other words, only when the bird is fully prepared does weather come into the equation.

When we review all the complicated internal preparations that are necessary for departure, we can begin to see how a bird depends completely on its inbuilt systems for this stage of the migratory adventure. This helps us to understand, too, how essential it is for the flight itself to be programmed.

## Finding the Destination

We have now established that a migratory bird has the correct amount of migratory restlessness or energy required to get it to its destination. You could say metaphorically that the migrant fixes one end of a piece of string of a certain length to its departure point, carries the string along, and when it reaches the end of the string it knows that it has arrived. So, providing the bird flies in the right direction (*see* Chapter 4), reaching its destination should in theory be automatic.

This certainly seems to be the case, except for one thing. What if the destination proves to be particularly unfruitful? What happens if a bird arrives in the correct place, but cannot after all be sustained there?

Well, having taken on board how a migration is pre-programmed from beginning to end, we must now wonder at its flexibility. As soon as a bird has completed its internal programme as described above, it is now 'unleashed' and may set off again to search the area for more suitable conditions. Many species become nomadic in the new regime of the wintering grounds, moving from place to place according to need. This does not just happen in the arid climate of southern Africa. Redwings and Fieldfares are also effectively nomads on their wintering grounds in Britain.

*Fieldfares – nomadic in winter.*

## Coming Back

Return migration differs in several important respects from outward migration. For a start, every practitioner has migrated before, including last year's juveniles, so at least some of the route is likely to be familiar. And secondly, most return migration takes place in a shorter time and probably in longer stages.

Despite these differences, there is little to suggest that there is any less internal control for return than for outward migration. In fact, return requires greater fat reserves, which can only be accumulated by the appropriate chemical and behavioural changes.

*One of the earliest birds to return in spring is the Northern Wheatear. This fresh-plumaged male has hastened north as fast as it can, and its reward will be a good territory and enhanced breeding success.*

## The Early Bird...

*In extreme cases, the juvenile of a very early migrant such as the Barred Warbler may begin its migration only three weeks after leaving the nest. Since some Barred Warblers only stay in the nest for 11 days after hatching, this means that a few individuals may make the transition from egg to migrant in just 32 days, at least in theory.*

## From Egg to Migrant in a Month

Migrant birds operate according to tight schedules. By moving to special sites for breeding, a migrant restricts itself to a short season of reproduction with a definite beginning and end. This means that, at the end of the breeding season, it is not just the adult birds that must get into condition for departure, but their offspring, too. The changing seasons will not wait for them; to survive they must leave at an appropriate time.

In effect this means that juvenile migrants have to grow up very quickly indeed. They must fledge, moult and then move out, all within the space of a few weeks. If a species typically migrates early it must develop even more rapidly. Juvenile Garden Warblers, departing in August, begin their moult when 20–30 days old, whereas juvenile Blackcaps, departing in September, wait until they are 40–60 days old. A similar accelerated development can also be seen in late broods of the same species. That same Blackcap's luxury of having 40-60 days before a moult only applies to first broods. If a young bird is from a second brood that does not hatch out until, for example, August, then it has about the same time as a young Garden Warbler to get into condition, which is very little. There is intriguing evidence that the accelerated development of second-brood youngsters actually begins in the egg.

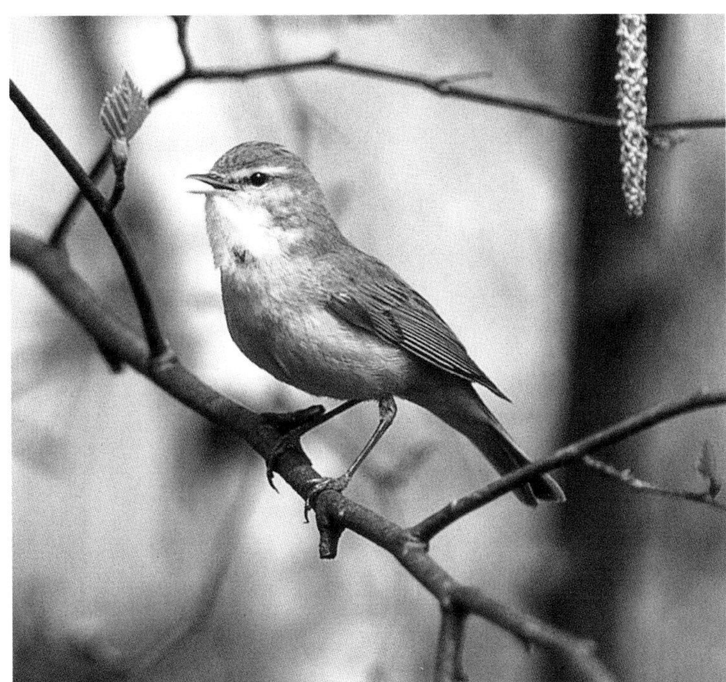

*Willow Warbler – the young have to grow up quickly.*

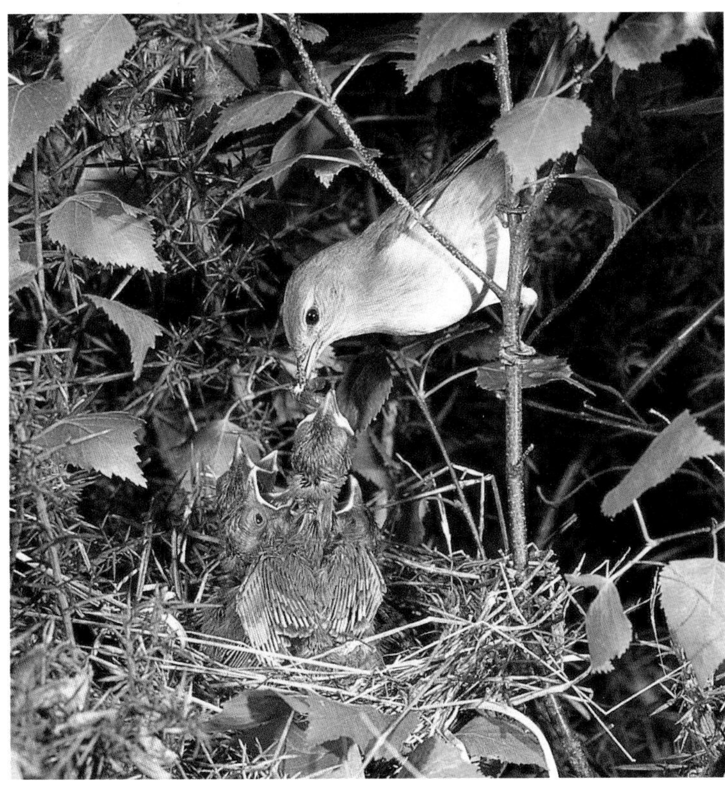

*The young Garden Warblers in this nest will be fully-fledged migrants within a few weeks.*

# 4 ORIENTATION MECHANISMS

*We've all heard stories of amazing feats of migration that serve to illustrate that many birds apparently have the capability to find a place with pinpoint accuracy from almost anywhere in the world. The next question is how do they do it?*

*Common Terns arrive in the UK in April. They begin their long journey to the west coast of Africa in August or September.*

## The Miracle of Bird Orientation

Here are a few accounts of remarkable migration journeys:

- A Manx Shearwater, experimentally whisked from its breeding grounds in Wales to the East Coast of the USA, managed to fly back to its own nesting burrow in only 15 days, travelling a distance of 4,000km in the process.
- The world distance record for a homing pigeon is 7,588km, achieved by a bird released on Guernsey in the Channel Islands in 1986. It made a transatlantic crossing and returned home to its loft in Brazil.
- Mourning Doves in the northern USA migrate south for varying distances, some travelling as far as Central America. A study in Minnesota found that 80 per cent of the population returned after their migration to within 61m of their breeding site in the previous year.
- A study of Spotted Redshanks in Finland proved that they consistently arrived between 1st and 8th May every year for 24 years.

*Manx Shearwater, famed for its long-distance migration.*

*In Finland, you can almost set your calendar by the arrival in spring of Spotted Redshanks.*

# What is Orientation?

When considering how birds find their way from place to place, it is important to sort out some definitions. In particular, we must distinguish between orientation and navigation. Part of the reason for this is that we know a great deal about orientation in bird migration and rather less about navigation. But we know for certain that birds can do both.

Navigation is the ability to find a specific location from somewhere else, whereas orientation is the ability to use an internal system to proceed in a certain direction. Navigation is accurate; orientation is approximate. Navigation takes a bird to a goal; orientation ensures that a bird sets off in a particular direction, and keeps it on course with reference to some kind of 'compass'.

One form of navigation is readily understood, and that is the use of landmarks. Many birds make short movements during the course of the day, perhaps commuting between a feeding site and a roosting site. It is known that birds can build up a picture of their surroundings over time, and may use this as a reference system to find their way around. In theory, a bird can utilize landmarks on any journey; for example, migrants probably use landmarks in the final stages of their return migration. There is nothing unusual about this – we also use landmarks to navigate around our local area.

There is an amusing aside to the use of landmarks. Recently, a study into the navigational expertise of carrier pigeons in southern England produced a startling conclusion. When making their way back to their lofts, the pigeons did not necessarily take the shortest route 'as the crow flies'. Instead, they preferred to follow

*Wasn't it the A31 we wanted?*

roads, turning off at the appropriate junctions and even tracking the appropriate twists and turns. Up in the sky, birds do not suffer from traffic lights or jams, so it seems that roads must be a blessing to lazy pigeons!

Looking out for landmarks is only part of the story. A young migrant bird faced with its first southbound journey to Africa cannot use landmarks it has never seen. Its technique must involve orientation, and this is where we begin to delve into the depths of a mystery that has intrigued mankind for centuries.

## An Internal Programme

In the Northern Hemisphere, Swallows fly south for the winter. They do not go north, west or east. It doesn't matter whether they are experienced migrants or birds recently out of the nest – they all travel in the same direction.

In various experiments using birds in cages, migrants with no external clues – no weather, for example, and no other birds to help them – fluttered in the expected direction at the right time of year. Young Garden Warblers that had been hand-reared and well-fed, and which therefore had no 'need' to migrate, also displayed the urge to move south in the autumn and north in spring. What can we make of such experiments? There is one inescapable conclusion. Birds 'know' in what direction to fly, and when. It is an innate, pre-programmed urge.

Birds' directional preferences are in fact finely-tuned. The Garden Warbler has a distinctive migratory pathway known as arched migration (*see* page 14) that takes it south-west from central Europe to the Iberian Peninsula, and then south-south-east towards Central Africa. So, midway during its migration this species has to make a distinct change in direction. Remarkably, Garden Warblers kept in cages in one place do indeed attempt to head south-west in September; later on, in October, their directional preference changes spontaneously to south-south-east.

*Garden Warbler – the precise details of its migratory path are pre-programmed.*

It follows that birds hatch with a pre-programmed migratory direction. This programme is connected to what must be a very accurate internal calendar, giving season-specific instructions. Such a calendar is not unusual among animals, and controls many other aspects of life, including annual breeding seasons and annual moult. It is usually calibrated to the changing length of daylight.

If we accept that a bird 'knows' that it should head south, for instance, then what reference cues does it use to maintain its heading? How does it know where south is? This is where the bird's internal 'compasses' come in.

*All migrants, including these Ring Ouzels, "know" internally which direction to take.*

## The Sun Compass

It is not difficult, even for human beings, to orient by use of the Sun. We all know that it rises in the east and sets in the west, and that at noon in the Northern Hemisphere it is in the south and also at its highest point in the sky. So it should be easy for birds to use this simple reference as well. But we forget one thing. Aside from the beginning and end of the day, the Sun is only a useful compass if you know what the

*The point of sunset is a useful reference cue for this Whinchat.*

*Starling, the first bird proved beyond doubt to use the sun in orientation.*

time is. Do birds have an internal clock accurate enough to use the Sun for navigation?

A few years ago, two experiments confirmed the existence of a **Sun compass** in the Eurasian Starling, a species that migrates by day. Firstly, it was found that Starlings in round **Registration Cages** (*see* page 99) would orient in the correct direction – that of wild migratory Starlings – only if the Sun was clearly visible, If it was cloudy, the birds' orientation was impaired. Then, by experimentally altering the apparent direction of the Sun using mirrors, the scientists found that they could alter the course taken by the caged Starlings. The birds indeed followed the bearing suggested by the false Sun. So the existence of a Sun compass in Starlings was proven. What about the internal clock?

Later experiments on pigeons demonstrated that, at least as far as the Sun's position could suggest it, birds do have a sense of time. The pigeons, also diurnal migrants, were kept in cages and subjected to an artificial regime of light and dark over a few days to shift their internal clock. When released into the fresh air and asked to navigate, they flew off in the wrong direction according to how much their internal clock had been altered. Many further experiments on all kinds of species have also demonstrated the existence of a highly efficient time-keeper ticking away in birds.

If birds use the Sun's position, with their internal body clock as calibration, this raises all sorts of questions. What happens when a long-distance migrant reaches and crosses the Equator, for instance, when the Sun's position changes? And how does the internal clock alter, not just to take account of shorter days as the bird approaches low latitudes, but also according to the seasons? Clearly, there is much work still to be done on this aspect of migration.

One might assume that a Sun compass is useful only to birds that migrate by day, and not to night migrants. However, this is not quite the case. Many nocturnal migrants begin their migratory flights not long after sunset, when they still have the lighter part of the sky to use as reference. And they can also detect something that we cannot see: polarized light. Light is polarized in a particular pattern across the sky according to the position of the Sun, and this pattern remains for some time after the Sun has set. Since birds can detect these patterns, it seems highly likely that they are used as another reference for orientation.

# The Star Compass

The remarkable discovery of the star compass in birds arose in a similar way to that of the Sun compass. First, there was an observation: birds seem to lose their way at night when there is heavy cloud cover. Then there was an elegant proof: birds in Registration Cages (*see* page 99) that could see the display of stars in a planetarium could be persuaded to switch the direction of their flutters if the stars were artificially flipped over by 180°.

It is not really surprising that stars should be used by birds. After all, stars are a dominant feature of the night sky, especially at the heights at which birds can fly. Moreover, stars move in a way that aids orientation – in the northern sky they appear to rotate about the Pole Star, which fixes north. Detailed experiments have shown that birds do not have to know where all the stars should be, but simply need to see a pattern of pin-pricks of light rotating about a fixed point. It doesn't matter what this pattern actually is, and once you edge away from 35° from the Pole Star the birds are immune to any alterations. Their star compass is, in fact, very simple.

*With so many birds migrating at night, it is perhaps not surprising that they use the stars for orientation.*

**NOT BY THE LIGHT OF THE MOON**

*There is so far no experimental proof that birds use that other celestial body, the Moon, in orientation. If anything, the Moon can be a bit of a hindrance to a migrant, presenting nothing more than distracting bright light.*

Although some birds clearly hatch with the ability to 'read' the stars, their star compass works only when they have had the opportunity to observe the night sky during the very early part of their lives. This has been demonstrated by the defective star compasses of birds that have been hand-reared and kept inside. It's quite a thought: imagine a young bird barely out of the nest, watching the night sky and working out its rotation! Presumably such a youngster would have to take a break from its ordinary nocturnal roosting to do this, or perhaps the learning phase coincides with the necessary switch in routine that immediately precedes a night migrant's departure.

## MAGNETIC FIELDS

Birds outperform humanity to such an extent in their orientation ability that scientists have always been on the lookout for a perception that birds possess and we do not. And in birds' **magnetic sensitivity** they found what they were looking for. Humans are

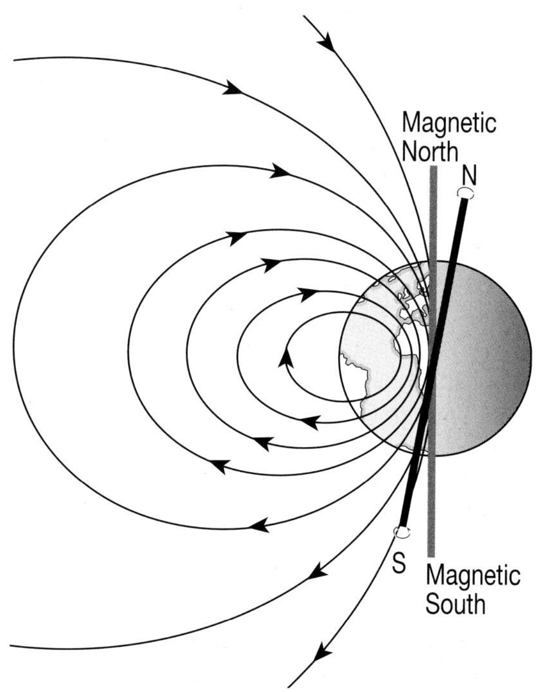


*A diagram of the magnetic field around the earth (left side). Note that the field lines run from south pole to north pole.*

*There's more to Robins than spades and mealworms; those that migrate can use the earth's magnetic field to help them orientate.*

basically immune to Earth's magnetic field, but birds – and many other animals – can detect it and potentially be informed by it.

Even before avian magnetic sensitivity was proven it had been suspected. After all, birds had to find their way somehow on dark, cloudy nights, without any visual clues to help them. The final proof came when Friedrich Merkel and Wolfgang Wiltschko put migratory Robins in an **Orientation Cage** cut off from external fields and, by means of huge electrical coils, altered the magnetic field within the cage. The Robins slavishly followed the predicted direction according to how the field was altered.

There are actually a number of different components of the Earth's magnetic field that birds could detect. They could detect its strength, for example, or monitor its direction (polarity). But the one most clearly proven so far is the angle of inclination of the **field lines**. Field lines run away from the magnetic South Pole and into the magnetic North Pole in such a way that they incline at 90° at both poles and run 'flat' (0°) at the Equator. In between they incline at progressively greater angles from the Equator to either pole, so birds, noting the change, can detect the part of the Earth's surface over which they are flying. They will, of course, also be able to tell when they pass the Equator itself, if only negatively, by not detecting any angle of inclination at all.

What physical or chemical detection mechanisms do birds possess? The Sun and star compasses rely on vision, and there must be something equivalent for the sense of magnetism. Research is providing more questions than answers – there does not appear to be any clearly defined organ that stands for the magnetic 'eye' or 'ear'. Some experiments have found that birds' magnetic sensitivity works only in certain types of light, suggesting that it is related to vision. But small amounts of magnetite, an iron-based magnetic substance, have also been found in the head of pigeons, between the skull and the brain and in some muscles of the neck. These crystals of magnetite are minute, and are found at different densities in different areas. It may be that they are fitted to tiny muscles and sensory nerve endings, which would register their collective movement in response to magnetic attraction and thus inform the bird about its magnetic orientation.

Whatever system they actually use, birds are remarkably sensitive to magnetic fields. They are, for example, often disoriented by magnetic storms and other magnetic disturbances, even sunspots. A magnetic storm only affects the total magnetic field strength at a certain point by a few tenths of a percent, so the effect is mathematically very small. But it is enough for the birds to pick up.

So we should not treat the magnetic compass as the least important of a bird's compasses, just because, perhaps, we cannot understand this sense for ourselves. On the contrary, the experiments so far suggest that it is probably the most reliable compass, and the one by which the others are calibrated.

## Smell

While on the theme of senses, what about smell? Could birds use their sense of smell for orientation? At first it seems improbable because birds have a much less well developed sense of smell than many other animals, including mammals, for instance. You just don't see birds going around sniffing. But there are those who argue strongly that the olfactory sense is not just important for migration, it is pivotal. The subject arouses much controversy in the world of migration studies.

Without going into the details of the argument, we should at least accept that there is a possibility that smell could be a factor, especially in local movements. After all, insects are drawn to pheromones over large distances. Salmon, too, reach their spawning

*Some seabirds like this Wilson's Petrel can follow a scent gradient to find food in the ocean.*

grounds by using an **olfactory gradient** as they swim upriver. Recently, it has been demonstrated that a small seabird, Wilson's Petrel, can detect the smell of chemicals given off by plankton and thereby identify rich feeding areas.

However, there is a big difference between locating a specific food and following a direction by means of scent patterns along the way. Orientation by smell would require a very sensitive nose and a sort of olfactory 'route map' informing the birds which smells are found where and in what concentration, and so on. It seems implausible, and for the moment, most scientists are sceptical about the importance of smell in bird orientation.

*Pigeons can hear sounds of low frequency inaudible to us.*

## Sound

There is another sense to consider, and that is sound. Can birds hear where they are going and adjust their route accordingly?

The idea would be fanciful had it not been for the discovery that birds can detect **infrasound** – that is, very low-pitched sounds inaudible to humans. Our audible range disappears at about 20Hz, whereas a pigeon can hear sounds of frequencies below 1Hz. That's another realm of experience completely. It is an exciting thought: at such frequencies a bird is able to hear the rumbling movements of oceans, the roar of high winds in the upper atmosphere, distant earthquakes, the wind disturbances around mountains, and all sorts of other geophysical clues.

Better still, as these sounds are of very low frequency they have a correspondingly long wavelength, enabling them to carry for hundreds of kilometres. With a little imagination we can appreciate the significance of this fact: can a bird in northern Europe hear the wind disturbances around the Alps and Pyrenees, or hear the rumble of the Mediterranean Sea and Atlantic Ocean? If so, it is not difficult to see how useful infrasounds could be in orientation.

One immediate problem with infrasounds is their potential susceptibility to swamping by similar sounds closer at hand. If a bird is being battered by high winds, for example, it probably cannot hear subtle, far-off clues. But the possibilities are there and, anyway, nobody is suggesting that birds can orient themselves purely on the basis of infrasound. It may simply be one cue among many.

*It is highly plausible that these incoming migrants heard the sound of waves breaking on shore long before land came into view.*

## Integrated Migration

By now one thing should be obvious from our discussion of bird orientation – there is no single, unified talent that enables birds to orientate in a required direction. Just as we might use several tools for one job, so birds do, too. Migration is an integrated process.

Research on one species, the carrier pigeon, or Rock Dove, has shown that it has a Sun compass, a magnetic compass and an accurate internal clock; it can also use landmarks, infrasounds and perhaps smell. It is also highly sensitive to barometric pressure, making it an excellent short-term weather forecaster. And the carrier pigeon is not even a true migrant!

We can assume, then, that every migrant has a range of instruments to help it in orientation. It might use one at a time or several at once. It is no less than a mini-computer, flying in the sky!

*In common with every migrant, the Little Ringed Plover will use a host of references cues to help it find its destination.*

## Navigation

However good a bird is at orientation, this is not the same as navigation. Orientation can take you in roughly the direction of your goal, but cannot pinpoint it accurately.

Navigation in birds is still a mystery. We simply do not know how a bird, displaced from one place to another artificially, is able to find its way back. It must work out its position of release relative to its goal, but the mechanisms of this are not known. It is an extremely exciting area of front-line research, and will doubtless lead to a sensational discovery in the future. To whet your appetite, speculation on this subject currently revolves around two main possibilities.

The first possibility is that birds have a map in their heads, which they use with reference to their different compasses; the map is informed by a host of clues, including visual landmarks, scent profiles, acoustic information, and so on.

The other possibility, which is particularly plausible in reference to the long-distance journeys that birds take, is that birds use a gradient system, using two constantly changing features of the Earth's surface in reference to each other. This can be understood by calling to mind the grid system that we use for map references: the measurements of 'easting' and 'northing' pinpoint a place with respect to how far east and north it is from two imaginary lines, the Equator and 0° longitude (the meridian). Birds cannot use imaginary lines, of course, but they could use two different features of magnetism, for example, such as the angle of inclination of the field lines at a particular place combined with the magnetic intensity or direction.

Or perhaps birds use maps *and* gradients? Who knows?

# 5 WEATHER AND BIRD MIGRATION

*It doesn't take too much mental gymnastics to realize that, since birds migrate in the fresh air, the timing and direction of their movements are heavily dependent on the weather. This assumed link has entered our culture. How many times have we heard that, if gulls come inland, there is going to be a storm? Or if unusual numbers of geese arrive at some location, it is a sure sign of a hard winter? Weather and bird movements are natural bedfellows.*

*The Snipe may make escape movements in harsh weather.*

## Folklore and Scientific Studies

Although most folklore about birds and weather turns out, with further investigation, to be false, there is undoubtedly a case of cause and effect at work here. Where most folklore gets it wrong is that bird movements do not normally predict the weather, they respond to it. Birds flee areas when thick snow covers their feeding sites, and they do indeed come quickly to ground when caught by inclement conditions while in transit. It is usually more sensible to assume that birds are the followers of the weather, not the other way around.

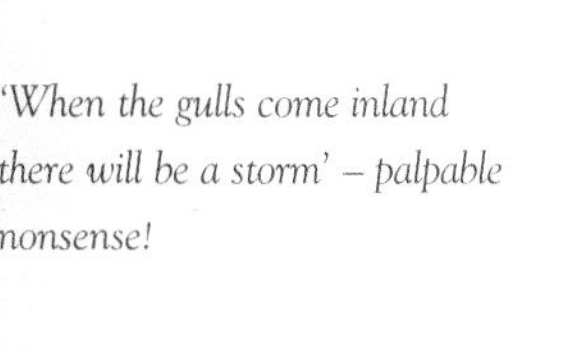

*'When the gulls come inland there will be a storm' – palpable nonsense!*

Having said that, recent studies on carrier pigeons and other birds have discovered that birds are able to detect very small changes in barometric pressure and can in theory respond to them. That makes them excellent readers of the atmosphere and potential weather forecasters, although only for the short term. A few birds can also potentially detect changes in weather by observing their food supply. For example, the number of flying insects in the sky begins to drop when a coming depression is still 500km away, perhaps alerting Swifts and other aerial feeders that lean times are approaching. However, the extent to which birds actually can or need to predict the weather is still uncertain.

Whatever birds' powers of prediction might be, their migratory movements undoubtedly go hand in hand with weather conditions. In the rest of this chapter we will take a closer look at this relationship.

# Suitable Weather

Which weather conditions favour migration? Two factors, it appears, are paramount: clear skies and a tailwind. Low temperature is not a problem; on the contrary, it can help to prevent birds from overheating during their flight.

To get the most from this section, we will need to refer to weather maps. They look complicated at first, but the most important things to remember about them are as follows:

- The wind direction around an **anticyclone** – also called an area of high pressure, or a high (H) – is clockwise.
- The wind direction around a **depression** – also called a low pressure area, a cyclone, or a low (L) – is anticlockwise.
- The closer the **isobars** (lines) around the weather systems are to each other, the stronger the wind.
- The weather fronts, which are marked with half-circles and/or triangles on the isobars, are generally associated with disturbances such as low cloud and rain.
- Any weather map is merely a freeze-frame of a dynamic situation. High and low pressure areas both move. Lows, especially, often well up from the Atlantic and move across Britain from south-west to north-east, and may pass in less than a day. By contrast, highs often remain in place for quite a while, and block the movement of low pressure systems.

*In common with all migrants, Swallows benefit from good weather when migrating.*

If we take on board that wind and clear skies are important factors in migration, we can expect birds to seek a tailwind for their flight and try to avoid upcoming fronts. Also, in the knowledge that weather systems move (albeit at variable rates), we can appreciate that birds sometimes need to exploit windows of opportunity that may be quite brief.

A weather map showing good conditions for outward migration, away from Europe in the autumn, is given on page 76. An anticyclone is centred over Britain, and it has all sorts of goodies associated with it: light winds, clear skies and no rain. What wind there is on the eastern side of the anticyclone is northerly, giving the birds a tailwind for their southbound journey. As a bonus, the northerly winds also bring cold air down, and so reduce the air temperature. An anticyclone like this may settle for several days, blocking the movement of air and creating ideal conditions for migrants. But go-ahead birds also take advantage of the somewhat stronger northerly winds that develop as a depression moves eastwards and an anticyclone moves in to replace it.

The situation on the weather map of good conditions for return migration in spring (*see* bottom map) is more or less the reverse of

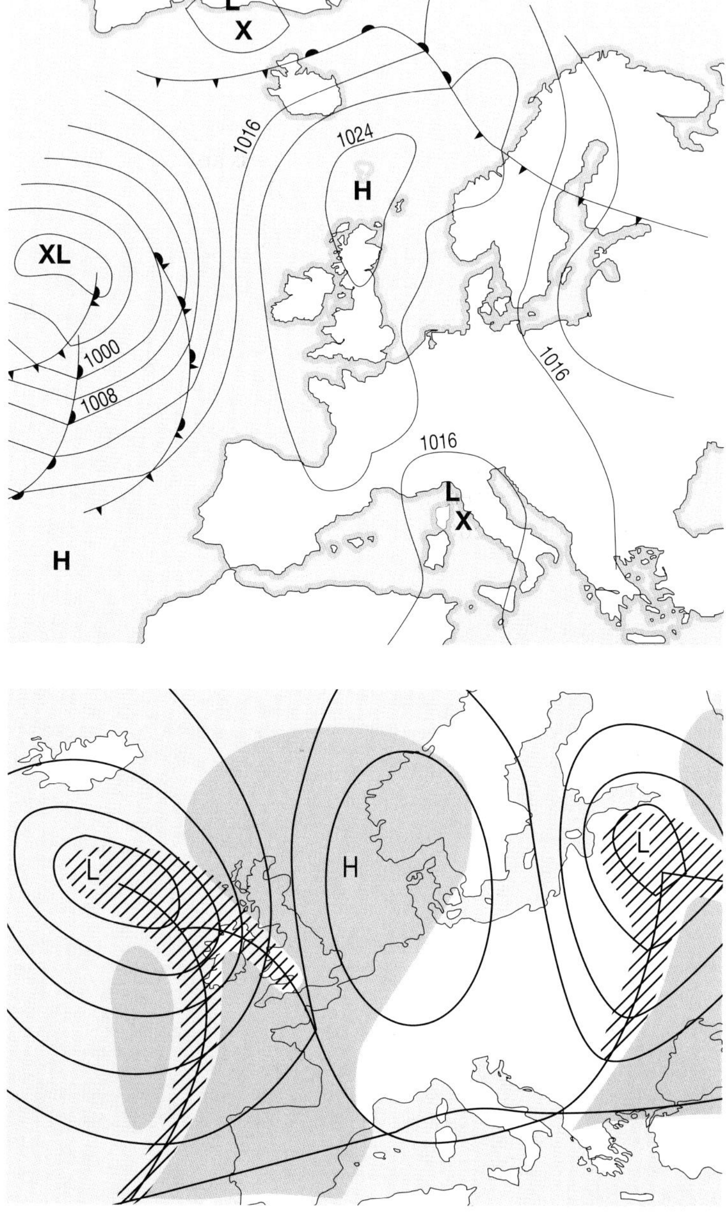


*Good weather for outward migration from Britain in autumn. The winds are light and the skies will be clear (well spaced isobars and no weather fronts). On the east side of Britain there might be a northerly breeze.*

*Good weather for return migration into Britain in the spring. In between a departing high and an approaching low, the winds are blowing strongly from the south and south-west. (Ideal areas are shaded).*

the outward migration situation. A high pressure area is departing east, to be replaced by a low pressure system from the west; in between them is a strong southerly or south-westerly airflow, the ideal tailwind for northbound migration. Incidentally, birds moving into this system are likely to find themselves amidst a warm front, and will often continue their journeys in moderate rain.

## Unsuitable Weather

If you take a look at the map below, you'll see every bird's worse nightmare for spring migration into Britain. In between a departing low and an arriving high are strong northerly winds, inhibiting the movement of birds. Such weather systems may play havoc with impatient migrants for weeks, but once they clear the travellers stream north as if some kind of dam has finally burst. Once the opposing wind has died down, things can really get going again.

*Black Terns often turn up in unexpected places in bad weather.*

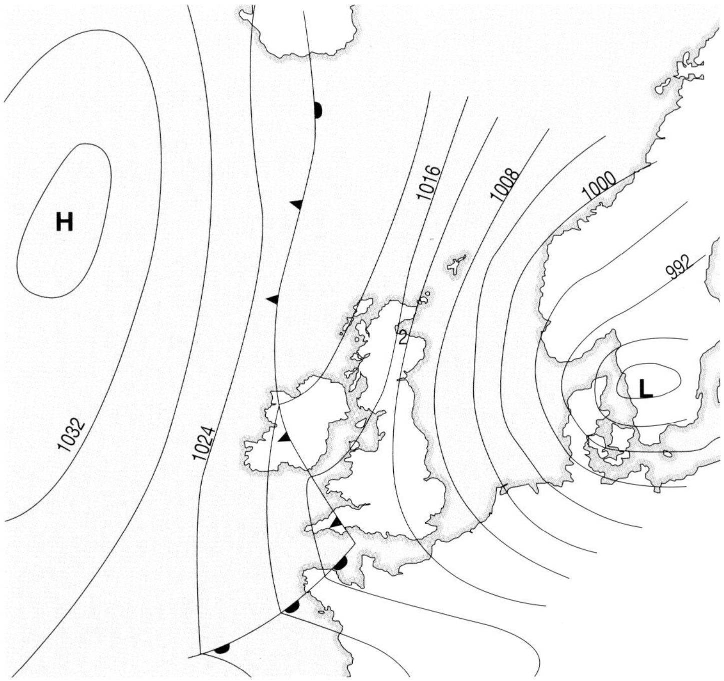

*Poor weather for northbound migration in spring. Britain is in between an anticyclone to the west and a depression to the east, resulting in northerly winds all over the country.*

Another, less common inhibitor for migration is, oddly enough, the absence of wind. One might suppose that no wind is better than a strong opposing wind, which is true, but one extra factor that we haven't considered yet is a side wind. Side winds, so long as they are not too strong, are useful. It is thought that they help birds to 'feel' the wind, and that the act of compensating for a side wind might actually help in orientation.

*A young Bluethroat in first winter plumage; young birds are the ones most likely to drift off course.*

However, it is only a short step from a useful side wind to a more troublesome wind. As mentioned at the start of this chapter, birds are utterly susceptible to the fickle nature of the air streams in which they are flying, and sometimes during the course of a flight things turn against them. One common hazard gives rise to what is known as **drift migration**. A bird sets out upon a specified course, using its various orientation mechanisms, and perhaps also assisted by a side wind. But then the side wind gathers strength, and the bird has some difficulty maintaining its course. It might make continual adjustments but, because of the difficult conditions, these adjustments are no longer sufficient to keep the bird on track. Eventually, it

*A 'fall' of migrants (Pied Flycatcher, Redstart and Wheatear) on a coastal headland. Their migration over the sea has been disrupted by cloud and rain.*

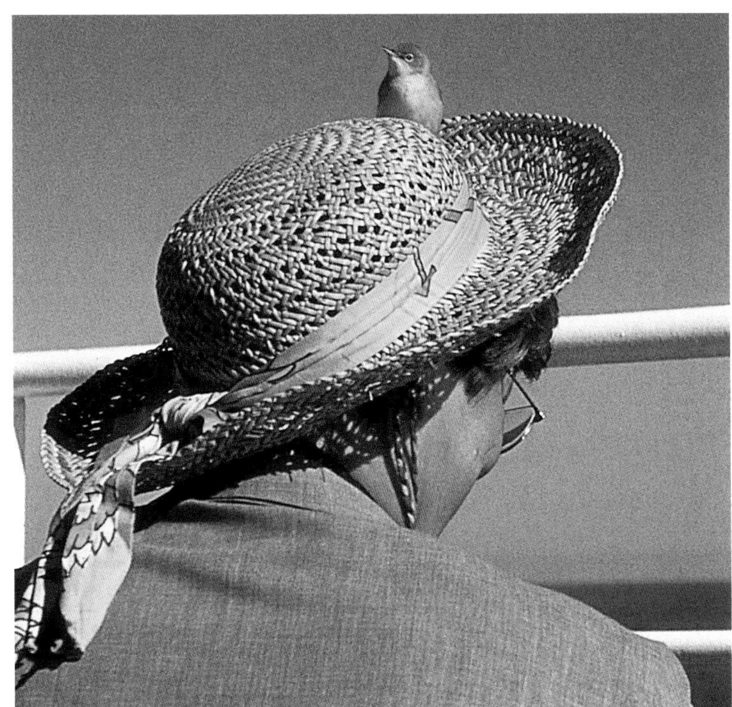

*'If this is a reedbed I'll eat my hat' – a Reed Warbler thoroughly misoriented.*

drifts further and further away from where it should be, and by the time it touches down, it is far outside its normal range.

Drift migration seems to happen most frequently to young birds on their inaugural flight, and the hopelessly disoriented creatures are therefore most often seen in the autumn. Their appearance often causes a stir among birdwatchers because a species outside its normal range is, by definition, a rare one. The excitement they arouse is, perhaps, one small compensation for the mini-tragedy of a lost bird. These hapless waifs do occasionally find their way back onto their normal track, but many more, one must assume, eventually perish.

## Emergency Touchdown

Another juxtaposition of circumstances also causes great excitement for birdwatchers. It occurs when birds set out during ideal conditions but then, suddenly, find themselves confronted with a weather problem *en route*. This could be a strong headwind, or heavy rain, or bad visibility, or any combination of the three. When this happens it is vital that the birds cease their migration as soon as possible. If they are flying overland, it might be only a passing inconvenience. If they are over the sea, it spells danger. In these circumstances, birds often land on oil-rigs or moving ships. More often than not, they will treat the very first land they

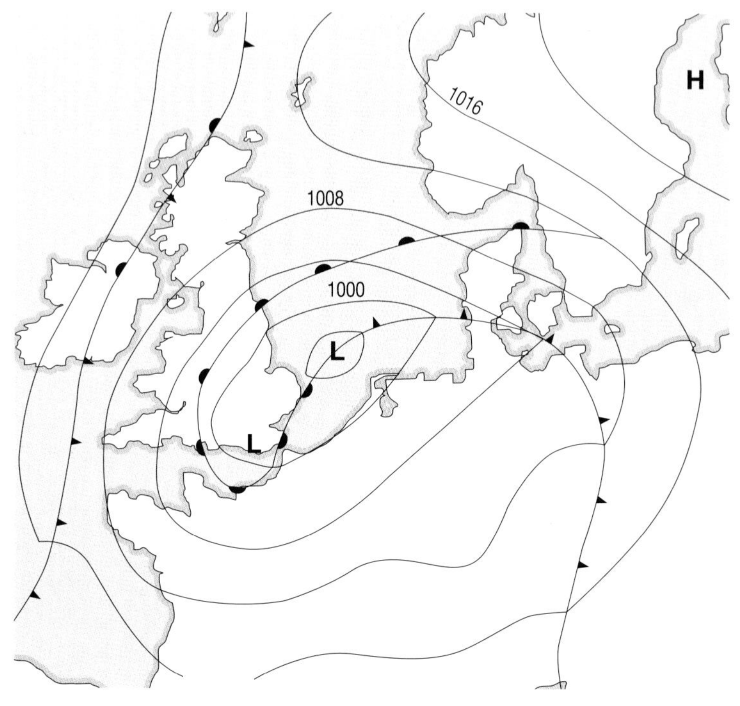


*Good conditions for a fall of migrants on the east coast of England. A low centred off East Anglia allows strong easterly winds to blow across from the Continent to the north, and the birds caught up in them meet a cold front upon arrival here.*

see as their long-lost home, and settle down immediately to sit out the bad weather. This explains why it is easier to see migratory birds in large numbers on the coast, and why certain geographical quirks such as islands or peninsulas seem to attract migrants like a magnet.

*Mist and rain may cause a 'fall' of migrants.*

It follows that a birdwatcher who loves seeing migrants ought to be an avid student of the weather. Depending on where you are, it can often be possible to predict a good **fall** of migrants – the situation when birds are suddenly grounded as described above. By way of example, *see* map above. A high pressure area over Scandinavia creates ideal conditions for departure (tailwinds, clear skies and no rain) but during the night a depression over central Europe moves north, shifting the relevant winds from north to east. Meanwhile, the accompanying cold front brings heavy skies and rain, immediately grounding the migrants upon their arrival in Britain. So it's definitely time for the birdwatcher to head to the east coast of England!

The same rules apply anywhere. If you want to see migrant birds, they must be aided by a tailwind heading towards you, and grounded by inclement conditions during the flight. A passing front will do the latter job for you. If you hope to witness spectacular migration during warm, settled weather, you will be disappointed: the birds will pass high overhead, invisible to the naked eye.

# Reverse Migration

When birdwatchers see a small migrant bird struggling to make landfall against the wind on a chilly day in the northern European spring, they often joke that the bird would be far more sensible to turn around and go straight back to Africa. They might not recognize quite so readily that the same option will have occurred to the bird as well.

*A Redstart caught in a spring snowfall. It might need to fly south again briefly to await a change in the weather.*

It will not fly all the way back to Africa, of course, but, if it has the opportunity, it will indeed retreat some distance to the south, perhaps a few tens or hundreds of kilometres, wherever it can find milder conditions. Such a movement is known as **reverse migration**, and it occurs frequently, mainly during the return movements of birds trying to reach their breeding sites quickly. In fact, it might happen several times in succession, the birds streaming forward on a warm front and then being forced back by a cold front, almost like multiple pitch invasions undertaken by confused fans before the end of a football match.

One might expect birds that have had to retreat once or twice to become a bit disoriented, especially when their internal programming is urging them to head north. But migratory birds, as ever, seem to have a solution. Apparently, when forced into reverse they retreat in linear mode, passing south down a specific axis. When moving north once more they retrace their steps and, if pushed back again, remain in that same imaginary groove.

Incidentally, reverse migration also occurs in autumn. Birds travelling south are disrupted by unfavourable weather, and return north again, aided by tailwinds taking them in the 'wrong' direction. Sometimes reverse migration in autumn takes migrants further north than their starting point!

*The tiny Storm Petrel survives storms at sea by keeping to the troughs of waves.*

## Storms at Sea

Seabirds are tough creatures. One can only wonder at the Storm Petrel, a tiny seabird with a passing resemblance to a House Martin – despite being tiny and lightweight, it somehow manages to cope with high winds and raging seas. Better than that, it thrives.

### Wind-blown

There are times when even seabirds are blown off course by exceptional winds. In western Europe, such winds are inevitably associated with fast-moving deep depressions that roll in across the Atlantic, and these can bring seabirds into contact with their alien habitat – land.

Usually, westerly gales have no worse effect than to bring seabirds within sight of the shore. The travellers can then adjust their bearings and find shelter somewhere along the coast. But sometimes the low makes landfall on a dark, cloudy, moonless night and the seabirds, unable to detect the change in landscape below them, are blown far inland. Many then settle upon the first water they see, perhaps a reservoir or ornamental lake, where they make an incongruous sight floating among the Mallards and Canada Geese. If a lot of seabirds suffer this fate, it is known as a **wreck**. Although some of the victims of a wreck may never see the oceans again, dying of exhaustion, observations suggest that these vagrants are quite good at recovering quickly and returning to the sea.

*Strong westerly gales may bring ocean-going Sabine's Gulls and Kittiwakes (bottom two birds) within sight of land.*

*Where am I? A Leach's Petrel forced inland by gale-force winds finds itself in unexpected company.*

*Although seabirds are well adapted to high winds and powerful seas, some do get blown accidentally inland by storms.*

## Escape Movements

So far we have looked at the effect of weather on birds that are already 'signed up' to migration – the individuals that have every intention of leaving one area and arriving at another. As far as these 'obligate migrants' are concerned the weather could be either a help or a hazard, but it will not be the *raison d'être* of their journey.

Some movements – you can call them migrations or not, it doesn't really matter – are actually induced by weather. They happen when, for example, birds wintering at a seeming favourable spot are forced to move on by a sudden deterioration in conditions. Perhaps snow and ice take a grip on the land, removing all access to food and water. The birds are then forced to flee for their lives and leave, one might say, without any warning and without time

Pink-footed Goose – a grazing bird that must escape frost and snow.

to pack their bags. They become refugees, taking part in **escape movements**.

Escape movements are actually quite common, especially in midwinter. Ducks, geese and grebes wintering in continental and northern Europe almost inevitably find their chosen waters freezing over at some point of the winter, so they transfer to the coast, or to the nearest free water. A change in weather may well see them back again soon. Then the next jolt of Arctic air sparks another evacuation. And so they go back and forth, like tennis balls in mid-rally, over the 'net' of the English Channel or North Sea.

Wildfowl are powerful fliers, but for other birds escape movements can be much more disruptive to their lives. Many smaller birds have to evacuate their winter quarters permanently – not just migrant birds, or even partial migrants, but residents, too. In Europe, escape movements tend to be directed both south and west. Britain receives its share of midwinter influxes, but southern France and Iberia take in the largest numbers of immigrants, which benefit from the relatively benign climate in these regions.

What sort of birds tend to be involved in escape movements? Water birds, with their dislike of ice, are obvious candidates. So are species that feed on the ground, especially those that search the soil for invertebrates or seeds. Lapwings and Skylarks are

*If conditions like this persist for long, birds will be forced to move*

*The first birds to escape frosty and snowy conditions tend to be ground-foraging birds such as Lapwings (above) and thrushes (left).*

often the first to evacuate, and they may be followed by gulls, thrushes, finches, buntings and even crows.

There are occasions when escape movements do not work for the birds concerned. In the harsh winter of 1962–63, for example, there were countless reports in Britain of birds literally dropping out of the sky, dead from exhaustion. Others perished almost as soon as they made landfall. The problem for many of these birds was that the freezing conditions were exceptionally prolonged; this meant that many birds that initially stayed to ride out the cold snap left when they were much weaker, when it was already too late. It is estimated that, during that fearful winter, almost half those birds remaining in Britain on Christmas Day were wiped out.

**Mass Movements**

*Woodpigeon escape movements can be truly spectacular – for example, the 30,610 that passed Landguard Point, Suffolk, on 2nd November 1994 – and so can those of Starlings – for example, 87,000 passed Hunstanton, Norfolk, on 16th October 1997.*

## Swifts and Weather

We might sympathize with migrant birds faced by high winds, squalls, heavy rain and snow – the ones that we can see touching down after their nightmare flights in a state of total exhaustion. But the Swift is not prepared to suffer in this way. It is the ultimate fair-weather bird, with the lowest tolerance of poor conditions that you can imagine; if it were human it would be the first on the picnic site to don an umbrella! At the merest drop in barometric pressure, the Swift flees, indulging in one of the most regular escape movements of any known bird.

There is a good reason for the Swift's intolerance: its specialized diet. The Swift subsists almost entirely on what is often termed 'aerial plankton', that is, the masses of insects and web-borne spiders that waft around in the air up to a few hundred metres

*Swifts feed on aerial insects in the layer above treetop height, and prefer warm, dry climates.*

above the ground. It takes just a small breath of wind to shift the aerial plankton around, and a spell of poor weather can reduce the levels of available food almost to nothing. So, rather than staying put and waiting for things to improve, the Swift abandons its normal summer quarters and flies in search of better weather.

## Swift Escape

The Swift often has forewarning of a coming depression. The amount of 'traffic' in the aerial plankton layer drops sharply well before a weather system arrives, and it is highly probable that the Swift has its own inbuilt sensitivity to barometric pressure (the carrier pigeon certainly does, *see* page 74). Whatever cues the Swift uses, however, it soon becomes aware of the need to take evasive action.

The nature of the Swift's escape movements is remarkable. We have all seen diagrams of cold weather fronts on TV or in newspapers, and if we have taken in nothing else about them we can at least appreciate that they are often very large systems; in fact, they can easily cover the whole of Britain. This is the scale of the Swift's problem.

Most depressions approach western Europe from the south-west, coming in from the Atlantic. You could describe them almost as waves, disrupting the weather as they pass. The aim of a Swift is to reach the better conditions behind the front, in the 'lee of the wave'. This often involves circumnavigating the depression, keeping away from the wet and windy middle and using the fairer winds on the perimeter to reach its goal. Since air circulates anti-clockwise around a depression, one might expect the Swift to head northwards over the top of the system, using tailwinds to help its journey – and perhaps many Swifts do. But, perversely, most field observations are of Swifts heading south-east at the approach of a depression, taking the short cut underneath it, against the wind. This explains why some British Swifts involved in these bad weather movements turn up in Germany or France.

After its long flight away from the brunt of the depression, a Swift might well have clocked up a round-trip of as much as 2,000km. By comparison with other bird migrations it is not much, but remember: this journey might have taken place in the middle of the breeding season. The birds involved may be adults with young, as well as young birds in their first full summer of life, not breeding yet but attending a colony. What kind of consequences might this behaviour have for the young in the nest?

*Swifts fly around depressions to avoid storms, even in the middle of the breeding season.*

Fortunately, young Swifts are tough creatures and have physiological adaptations to cope with periods of starvation. For one thing, they lay down extra fat reserves almost from hatching, in case of lean times. They are also able to enter into a state of torpor, reducing their metabolic rate to save energy until the next meal is delivered. After the parent Swifts' long return flight in fair weather, one hopes that their first delivery of aerial plankton to the hungry youngsters is a sumptuous one indeed.

## Rare Birds and 'Accidentals'

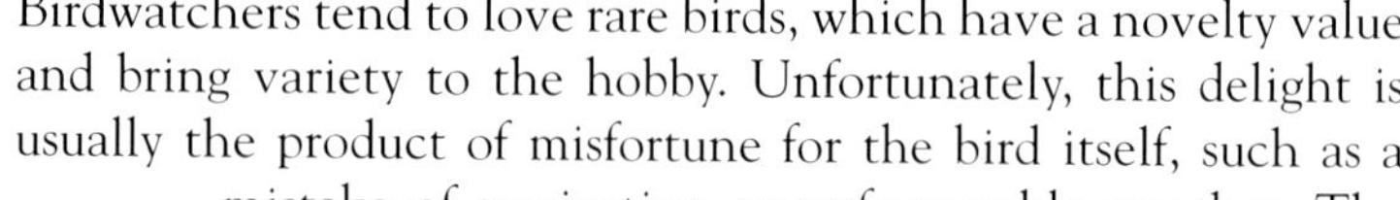

Birdwatchers tend to love rare birds, which have a novelty value and bring variety to the hobby. Unfortunately, this delight is usually the product of misfortune for the bird itself, such as a mistake of navigation or unfavourable weather. The truly rare birds have made the greatest mistakes, ending up thousands of kilometres from where they should be. But birdwatchers shouldn't feel shame for admiring rarities or benefiting from these mites' misfortune. After all, the birders are not responsible. If you wish to be a bit mercenary about it, the rarities are probably members of that pool of birds destined to die before reproducing – the so-called 'doomed surplus'.

*It takes an exceptional quirk of weather to bring the American Blackburnian Warbler to Europe.*

In Europe, we have two main categories of extremely rare bird: those that have come from the Americas and those straying from Asia. We get the occasional out-of-place seabird or Arctic stray, and prolonged southerlies occasionally bring birds from the south, as **overshoots**. But the most exciting and most regular arrivals come from the west or east.

*The exotic Bee-eater regularly overshoots its normal breeding range to turn up in Britain.*

Despite their rarity, some of these unexpected visitors show a pattern of occurrence that can be traced back to weather systems. For example, transatlantic vagrants usually turn up in the wake of a succession of fast-moving Atlantic depressions. A good many species travelling between North and South America do not follow the coastline but fly out over the western Atlantic because it is quicker (*see* page 112). If these birds encounter cold fronts and associated rain and wind they become disoriented. Then it is all too easy for them to get caught up in the conveyor belt of Atlantic lows heading north-east.

It is less easy to explain the arrivals of birds from Asia. Sometimes huge high pressure areas form near their breeding grounds and nudge them westwards, at least enough to be transferred, baton-like, to a depression that will whisk them anticlockwise down towards Europe. However, these strays do not always arrive under such conditions, and other alternatives, such as disorientation on a grand scale, should be sought.

*The rarest birds attract many admirers. Birdwatchers that spend most of their time chasing rare birds are called 'twitchers'.*

## The Mystery of the Pallas's Warbler

*Pallas's Warbler and other Asian warblers appear to be becoming more and more regular in Britain and Europe, especially in autumn – so much so that the weather is no longer being used as an explanation for their appearance. One highly intriguing alternative theory is that the birds are breeding in an area of magnetic disturbance, which disrupts the orientation of the juveniles. Such an area does exist in Asia, where the mountains are rich in iron ore. Could this explain the mass mis-migration?*

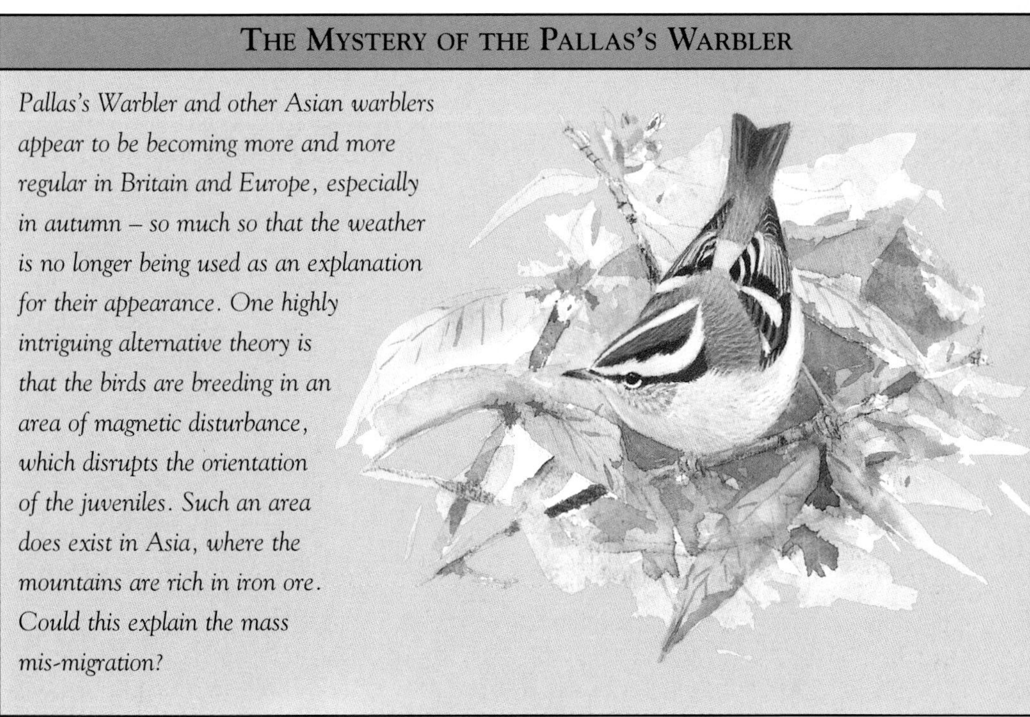

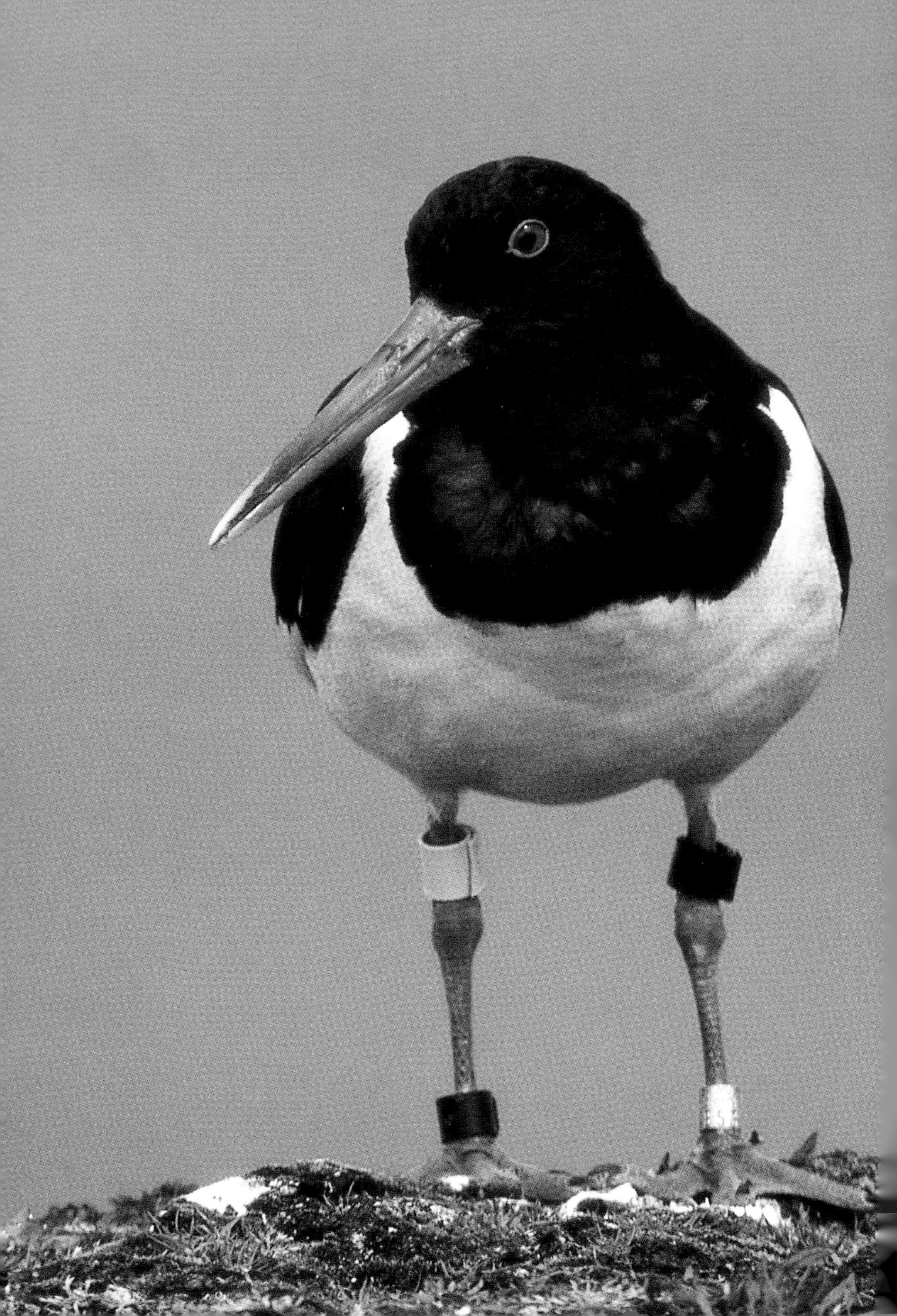

# 6
# STUDYING MIGRATION

*Anyone can study migration – you can count the number of Swallows or Swifts passing overhead, for instance, and observe how that number varies by day and according to the weather. Interesting results can also come from recording what species are found in a certain location from day to day and month by month. This proves that birds move around. Direct observation is the simplest form of migratory study, to which anyone can contribute.*

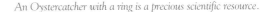

*An Oystercatcher with a ring is a precious scientific resource.*

## Ringing and Trapping

Observing migration in the field has one major drawback. It highlights only species, not individuals, unless the latter are colour-marked (*see* below). Thus we might observe a Willow Warbler in Finland in April and another Willow Warbler in Senegal in November, but we cannot be sure that it is the same one unless it is fitted with some kind of unique tag. This is where ringing comes in. Ever since the beginning of the 20th century, when a Dane called Hans Christian Mortensen cut the first metal rings for birds, ringing has been in the forefront of research on bird migration.

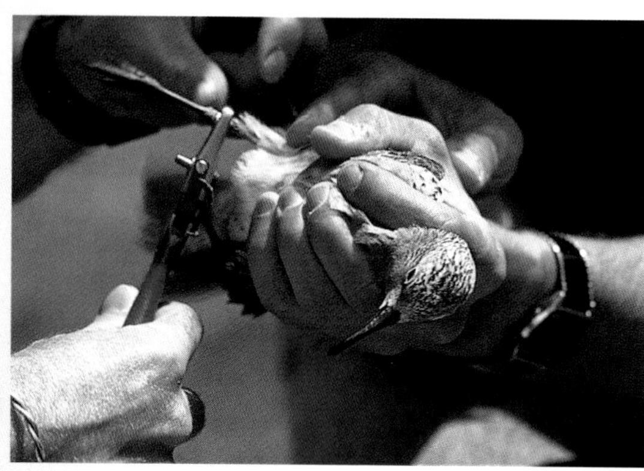

*The act of ringing can be stressful for a bird, but there is no evidence that rings are uncomfortable or impede movement.*

Ringing – or 'banding' as it is known in the USA – has come a long way. Nowadays there are almost 20 different kinds of ring of 2–26mm in diameter, tailored to suit the size of the bird. They are usually made of aluminium or some other lightweight metal, and are designed to last a bird's lifetime. They fit comfortably on the most scaly part of a bird's leg and cause no damage by rubbing. As far as is known, they do not cause any discomfort: birds avoid predators, go to roost, moult, copulate, incubate eggs, feed young, bathe and migrate – all with a ring fitted. Indeed, with their sophisticated design and personalized number, rings really ought to be the height of bird fashion!

The unique number embossed on a ring is effectively a PIN number and every ring also carries an address. The address that has been standard for British rings since 1909 is 'British Museum, London', and to prevent confusion every other country also has its own central monitoring scheme with a single address. When a dead bird is picked up, or if it is caught again for ringing, the ring or information can then be sent back to its place of origin. (If someone finds a dead bird and returns the ring, they in turn receive an information sheet detailing the known history of the individual bird concerned.)

*A selection of different sized rings.*

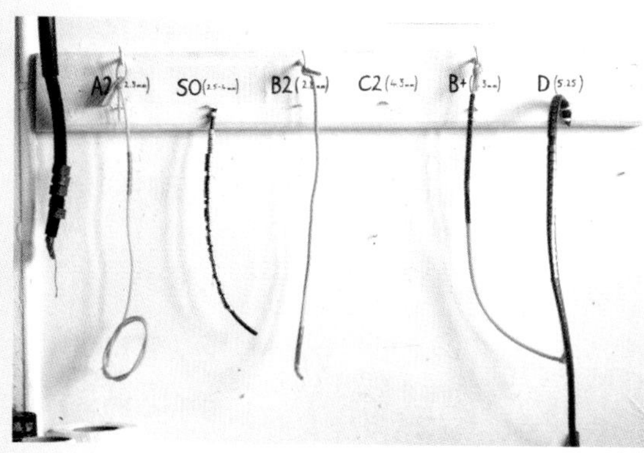


There are two serious drawbacks to ringing. The first is the exceptionally low recovery rate compared to the enormous effort that is made by ringers. Small birds such as warblers may have their rings returned at a rate of less than one bird in a thousand, although larger and regularly

*If you find a dead bird like this Reed Bunting, remove the ring and send it to 'British Museum, London'.*

hunted birds such as wildfowl are recovered much more often – up to 20 per cent of ringed individuals. Nevertheless, every ring is precious, because it may only take one successful ring recovery for us to be able to understand a link in the chain of a bird's journey.

## Traps and Nets

The other serious drawback to ringing is that the birds have to be caught first before they can be processed. This undoubtedly causes considerable trauma to a bird, and might potentially also affect its overall condition. Occasionally birds are killed in nets, mainly by predators that cunningly have learned to exploit this unusual source of easy food. But this drawback must be seen in the light of the priceless information base that ringing provides, including important evidence to support conservation efforts. Ringing is also carried out with almost paranoid care, so that the death of a bird in a ringer's hand is a very rare event.

You cannot become a ringer overnight. It is a long and laborious process and, effectively, you have to take out an apprenticeship with an experienced practitioner. To obtain your 'A' licence, which enables you to undertake ringing on your own, takes about five years. This is just as well because ringing requires considerable practice and quite a lot of learning. It is also a serious scientific tool and should not be taken lightly. It would be pointless to start ringing birds only to abandon it later and take up another pastime. Many bird ringing schemes have been in operation for many years, and often they are providing the most valuable information on changing bird populations.

There are various ways to catch birds, but the most successful and widely used is the **mist net**. This is made of thin nylon threads, making it difficult to see, especially in calm conditions, and few birds are sharp-eyed enough to avoid it. The shape of a mist net is roughly rectangular and for maximum portability it is erected like a tent, with poles and guy ropes. There are taut strings at different heights along its length; in between these the mesh billows slightly, so that when the bird hits the net it does not bounce off,

but instead falls into the 'pocket' so created. Almost anything from a Goldcrest to a Sparrowhawk can be caught in a mist net although, given the choice, most ringers would probably prefer the job of extricating the former!

A more permanent type of trap is the ingenious **Heligoland Trap**, developed on the island of that name in Germany. It is effectively a funnel-shaped cage, open at one end and with a box at the other. It is built into the ground and is therefore used at sites where ringing is carried out all the time. The trap's open side usually houses various bushes, which migrants love. When a few birds are deemed to be present inside the trap, several people then walk inside in a line, scaring the birds towards the narrow end. The birds dive into the box and are caught.

Ringers check their nets or traps every half an hour or so to disentangle their charges, and the birds are then transferred to thick, soft bags. The birds are processed immediately. Each one of them in turn is aged, sexed and weighed, and various measurements, such as wing-length, are also taken. The ring is fitted using a special pair of pliers, and the bird is released after its brief ordeal.

*Mist nets were first used to trap birds in the UK in 1956.*

*The Heligoland trap was invented in Heligoland, an island 50km off the northwest German coast. Heligoland is home to the world's oldest bird observatory, with records dating back to 1840.*

Anyone can watch ringing take place, so long as they do not touch the birds. Most ringing groups begin their sessions at first light, when the night migrants touch down, and continue into mid-morning. Spectators are welcome, especially at bird observatories (*see* page109), and are often invited to do a round of the

nets. This is the point at which you first see the wriggling birds and find out if you've caught anything extra-special. It is a nerve-tingling experience, and well worth the price of an early start.

It is estimated that, over the years, more than 200 million birds have been ringed around the world, and that figure continues to rise sharply. However sophisticated other techniques for studying bird migration become, the humble ring is likely to keep its importance for many years to come.

*Coloured rings may identify a bird individually in the field, without it needing to be caught. This is a Mediterranean Gull.*

## Other Types of Marking

There are other ways of marking birds apart from putting small metal rings on their legs. For a start, you can put larger plastic ones on their legs, several at a time, and in different colour combinations. In this way, larger species can be fitted with a unique, individually recognizable combination of bands that can be seen and recorded in the field.

A more drastic procedure is to dye the birds. A partially yellow gull among a flock of white gulls is easy to see, so this technique has the added advantage of making its objects conspicuous. In fact, some reports of extreme rarities have turned out to be birds marked by researchers! Dyed plumage might have a questionable effect on the birds, possibly increasing their vulnerability to predators and compromising their allure to potential mates, but as yet there is no proof of this.

*The dye on this Black-headed Gull could identify it as being from a specific place.*

*A wing-tagged Red Kite. This graceful bird has recently been successfully re-introduced to England and Scotland.*

Plastic wing-tags are also used in some studies, especially those concerning birds of prey. These tags are conspicuous and easily recognizable, and are highly unlikely to have any harmful effects.

The advantage of all these techniques over ringing is that the birds can be recognized in the field without being recaptured or disturbed in any other way. This tends to lead to more 'recoveries' – observations of marked birds. Some individual wildfowl have been sighted and recorded more than two hundred times, thereby contributing a great deal of useful information.

## Radar Studies

*Cranes were ideal subjects for early experiments with radar transmitters, as they are large and powerful fliers.*

One of the main problems affecting the study of migratory birds is that most of them travel by night. You cannot, therefore, track them by eye when they are moving. But, with the invention of radar during World War II, the hitherto secret world of nocturnal migration was uncovered at last. Using radar you can 'see' birds travelling – at any time of night, at any height, in any weather conditions and over a very wide area.

The word 'radar' stands for Radio Detecting and Ranging, and the technology works by sending incredibly rapid pulses of radio waves towards an object in the air. The rapid bursts reflect back and are detected as echoes. The radio waves always travel at exactly the same speed (the speed of light), so the time they take to echo back will be fractionally – but measurably – different according to how far away the object is. In recent times radar has been developed so that individual objects or groups of objects can be followed as they move through the sky.

Radar studies are of enormous importance in bird migration studies, especially since they catch birds 'in the act', so to speak. Radar can monitor a flying bird's height, direction and speed. It can reveal thousands of birds on a single screen – they look like clouds – and measure the density of migrants in the sky.

All of this would be perfect except for one thing. Unless a bird has a particularly distinctive way of flying – the Swift has recognizable wing-beats, for example – radar cannot distinguish between species, only rough groupings such as bird orders or families. Nonetheless radar is still highly instructive, and will remain one of the major tools for studying birds.

*A small transmitter being fitted on to a bird* (below).

*Whooper Swan with Radio transmitter* (bottom).

## Tracking by Radio and Satellite

The tracking of individual birds by radio waves is a comparatively recent experimental technique, but its possibilities make students of migration positively drool with anticipation. Already the research projects are proliferating, especially those using satellites. The thrill is that you can follow a single bird, individually, throughout the course of its migration.

The first experiments of this kind involved radio transmitters and essentially ground-based receivers. Devices fixed to birds transmitted radio waves that could be picked up as audible bleeps on a portable directional aerial. The birds could then be

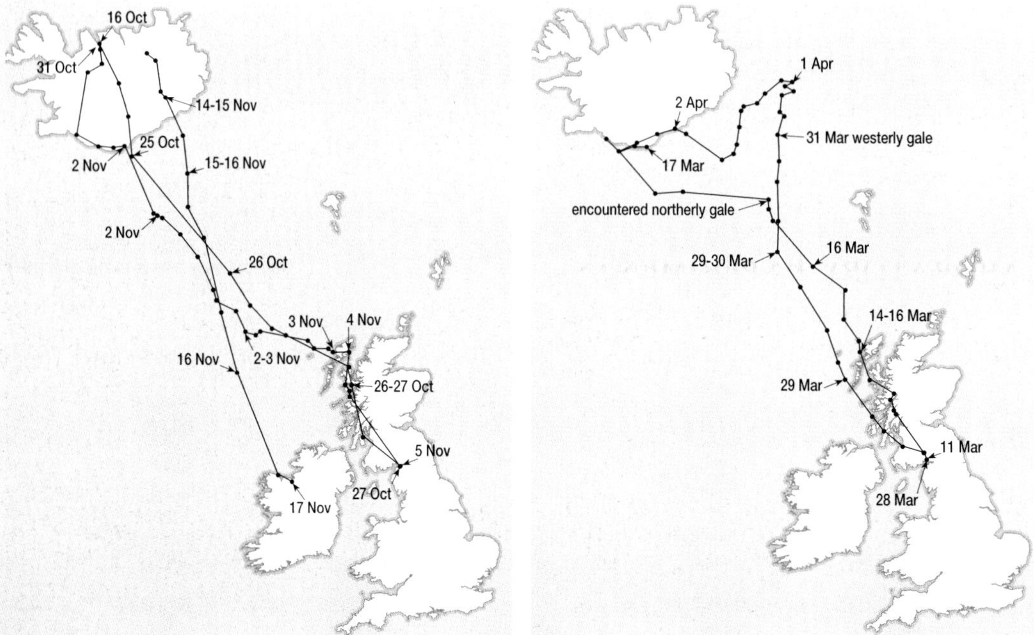


*The actual flight paths of Whooper Swans satellite tracked on the migration between Iceland and the British Isles (left, outward migration; right, return migration).*

followed by foot or vehicle, and their movements tracked. At first the transmitters were so unwieldy that only the very largest species, such as swans and storks, could wear them as 'knapsacks' on their backs, but they have become much smaller and widely used, even fitting small birds such as warblers.

These sorts of transmitters have always suffered from the dual problem of limited range and lifetime, which means that if the signal disappears the researcher is unsure whether the subject has strayed out of contact or whether the device's battery has run out. There is also a problem of cost and resources: following a single bird can involve a lot of people and require much energetic driving around or walking. However, this has not rendered radio transmitters invalid, and their use continues to this day.

A slightly different version of this type of study is to fit birds with data loggers. These record various details of the bird's movements and physiology, and build up a store of information equivalent to an aeroplane's black box flight recorder. An obvious drawback is that the bird must be recaptured for the data to be analysed. This can be circumvented by injecting tiny transponders under the bird's skin and recovering the data through nearby sensors, but as yet the range of such sensors is far too short to contribute much information on a bird's movements.

On the other hand, satellite tracking has a global range. The principle is exactly the same as for the other techniques just described – the birds are fitted with radio transmitters – but the information is conveyed to a satellite rather than to a ground-based receiver. This system has created, for the first time, the opportunity to follow a bird at daily or hourly intervals for the whole of its migratory journey, and even some of its daily activities can be monitored, too.

**The Latest Technology**

*Satellite tracking has a big future in studies of migration. It is the reality version of study, where researchers can virtually accompany the bird on any leg of its journey, anywhere in the world. Several such studies have their own websites, allowing a truly global audience to follow developments step by step.*

## Laboratory Experiments

Many of the research findings mentioned in this book have been made in the laboratory. It would be pointless to describe the nature of each study because, in a sense, every experiment is different and requires its own specialist equipment.

However, I shall make mention of just one such apparatus, because it has contributed so much to the study of orientation. This is the so-called Registration Cage, already mentioned in Chapter 4. There are several types, each designed to test a bird's preferred direction of fluttering or hopping, in the presence of absence of the orientation clue being tested. The cages are all circular, so that a bird is not inhibited in hopping towards any particular direction and moves purely on the basis of its migratory drives.

One form of Registration Cage is narrow at the bottom and wide at the top, so that a bird has to shuffle or flutter 'uphill', so to speak, towards its preferred direction; its movements register as scratches on correction paper. Another type provides ample perches, each of which give slightly under the weight of a landing bird, and trip a switch. The switches record how often, and in which direction, a bird moves around.

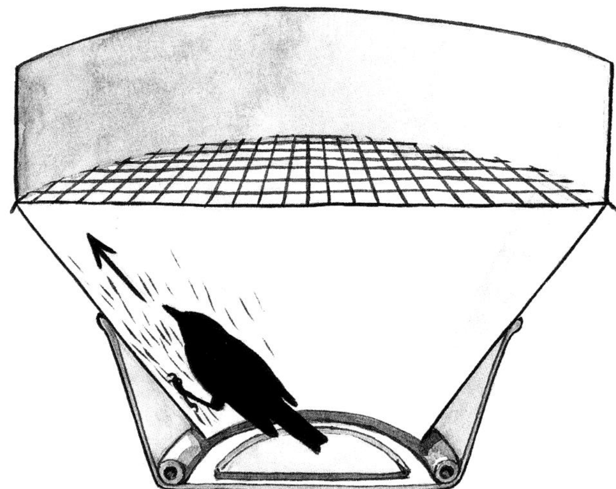

*The Registration Cage measures the direction in which a captive bird flutters.*

# 7 OBSERVING MIGRATION

*We've heard about it, read about it, but have we actually seen bird migration in action? If not, this chapter gives an idea of how to catch up with birds while they are on their travels. In this section we will not be concerned so much with watching the results of a migratory movement (i.e. what species have turned up at a place overnight), but with being a spectator and witness to the actual event itself.*

*Birdwatcher at dusk.*

*Anyone can observe House Martins migrating over their house or garden.*

## SEASONS

Although migration of one type or another will be underway at any given time of year, there is no doubt that certain seasons experience heavier migratory traffic than others. In western Europe the two main migration seasons are the spring and autumn. This is because many millions of birds have to evacuate our area to spend the winter in warmer climates. So the autumn sees their departure and the spring sees their arrival. A good many birds also move within Europe, perhaps departing high latitudes for lower ones; since these movements have the same purpose as the longer journeys, they take place at about the same time.

The birdwatcher hoping to witness migration is therefore most likely to strike lucky between March and May, or August and November. The autumn migration is more protracted than spring migration because more birds are involved – adults plus the young produced that year – and they are not in such a hurry to get to their destination.

*The first Chiffchaff often heralds the start of the spring migration.*

# Visible Migration

Nightingale – its song can only be heard between April and June, and then it departs early, in August.

Apart from the special case of watching the Moon (*see* page 108), it is generally almost impossible to observe many migrants in action, because they fly during the night. However, you can often catch their landfall at dawn if you wish to make the effort. You might not see much, just a few shapes coming in off the sea or down from the sky, landing for a moment or two and then moving on. You might also hear new arrivals calling in the half-light. But watching **visible migration** gives the birder a very special buzz, especially since there is little way of predicting what bird will make landfall when. This is the 'lucky dip' of bird migration.

In difficult conditions such as rain, high wind or fog, birds' migration may be delayed, and so arrivals can be seen later in the morning. This may also happen if the birds have just completed a particularly long migratory leg that has taken more than a night's flying. Then you will see many more birds as they arrive. But it is hard to predict when and where such movements will take place.

If the skies are clear and the winds are favourable at the end of the day, you can also gain an impression of the departure of night migrants. This is the case almost anywhere at dusk. You may just notice a few hints of movement or calls, or perhaps even a flock of wildfowl rising excitedly into the air *en masse*.

*Starlings are among the more easily observed migrants.*

## Day Flights

However, the fact is that it is always easier to see diurnal migrants in action. This is partly for the obvious reason that they are not enshrouded in darkness and partly because they tend not to fly as high in the sky as night migrants. They often fly low to make use of any irregularity in the land to afford a more sheltered flight. They also tend to follow so-called **leading lines** – coastlines, rivers, mountain ridges and valleys – and this means that their movements can be predictable and may be concentrated. The most spectacular migrations are always along some form of leading line. Interestingly, diurnal migrants are also more liable to fly into a headwind than their nocturnal counterparts.

*Common Scoters tend to follow the coast when migrating, using it as a 'leading line'.*

*Whimbrel following the coast.*

In common with nocturnal migration, the actual journey of a daytime migrant often begins in the half-light, this time at dawn. Perhaps this enables birds to escape the early attentions of predators as they set off. The movements then continue for variable lengths of time, depending on the species involved. For example, scientists have found that Dunnocks migrating through a pass in the Alps travelled for the four hours after sunrise, and then numbers tailed off; this is a very common pattern, so an early start is often best to see visible migration. Swallows and Swifts are an exception and carry on intermittently throughout the day. Birds of prey are mainly seen from mid-morning onwards, until the thermals stop forming in late afternoon. They are particularly likely to be seen after a clear night.

In Britain, seasonal diurnal migration is best seen along coasts. Almost any coast will do and, if there is an onshore wind, birds such as waders and seaducks as well as various seabirds should be seen. Small birds also use coasts, of course, and they may be seen flying along beaches, or over headlands and cliffs.

*A birdwatcher 'seawatching'.*

But you do not have to go to the coast to see migrant birds flying by day. A large body of water, such as a lake or reservoir, is often a magnet for travellers. A hilltop can be good, too, especially one that affords good visibility in all directions. The trick is to find the most appropriate local site for you. Other birdwatchers will probably know of one, or you could follow your nose and locate one for yourself.

*River valleys and scrubby areas are ideal inland sites for watching migration.*

## Birds of Prey

Watching migrating birds of prey is a specialist pastime. Although many birds of prey are certainly migratory, in Britain they move along a broad front and it is difficult to see them in any sort of numbers. That said, if you select a narrow strip of land jutting out to sea, and wait for fair winds and mid-morning Sun at the appropriate season (such as September), you might well be lucky. You might also strike lucky if you watch along a mountain ridge or other high ground.

In this country when watching migrating birds of prey it is often beneficial to have both a light breeze and some cumulus cloud. Try not to be distracted by bird activity on land and instead keep your eyes fixed up on the skies. Birds of prey have this wonderful habit of simply materializing: one moment there is nothing, the next reveals several soaring dots. Sunglasses are often helpful to pick them out.

Birds of prey rely partly on thermals to migrate (*see* pages 41, 115–116), which means that the most intense movements are around the middle of the day, the peak time for rising currents of air. But thermals do not form over water. Most species of raptor, therefore, have to adapt their migratory pathways to avoid long sea crossings or other hazards (apparently, the Honey Buzzard is an exception). This concentrates them at suitable places in vast numbers.

*Above: Many birds of prey are large, heavy and broad-winged. They need thermal updrafts to help them migrate, so they avoid long sea-crossings. (clockwise from top: Sea Eagle, Rough-legged Buzzard, Golden Eagle, Honey Buzzard).*

*Right: Mountain ridges often concentrate migrating birds of prey.*

Britain has no such hotspots – after all, we are an island and a sea crossing is required to get here – so the raptor enthusiast has to travel to find them. The three best in Europe are Falsterbö in southern Sweden, where birds funnel down from Scandinavia, and Gibraltar and the Bosphorus at either end of the Mediterranean. At all these sites it is possible to see thousands of birds of prey every day. Other good areas include various locations in Israel and Jordan. The best sites in the USA include Hawk Mountain, Pennsylvania, and Cape May, New Jersey.

## Seawatching

Some of the world's longest migrations are performed by seabirds (*see* pages 112, 118), but their routes tend to follow a track far out of sight of land – and birdwatchers. In certain weather conditions, however, even the most ocean-going species such as various storm petrels and shearwaters may be forced reluctantly towards coastlines. Then, given a telescope and a lot of luck, it is possible

*Seawatching is a demanding form of birdwatching, requiring both skill and persistence.*

*A scene from every seawatcher's day-dream: top to bottom, Arctic Skua chasing Common Tern, five Manx Shearwaters, Great Skua, Great Shearwater and three Storm Petrels.*

### Taking Your Time

*Perhaps more than any other form of birdwatching, seawatching requires patience. You might have to wait many minutes, or hours, between birds. So it does not suit everyone.*

to engineer a heart-stopping – if brief – appointment with one of these displaced birds. This is the main aim of seawatching.

Not all seabirds are so disdainful of land. Many, such as gulls and terns, actually use the coast to guide them when migrating, and these species, too, are seawatching targets. Actually, if you give the term 'seawatching' its broadest possible meaning, it could encompass anything seen by looking out to sea, even travelling songbirds. But traditionally, the purpose of seawatching is to watch seabirds only: the rest comes under the remit of 'visible migration'.

Successful seawatching is highly weather dependent and wedded entirely to local factors. Apart from the rather obvious fact that you will not see much if there is a strong wind blowing offshore, which will blow the birds away from you, it can be hard to predict the best conditions. Ask a local sage: every sea-watching site has them! Do not necessarily expect the best results during a gale force wind – your prospects will be determined by what has been happening during the previous hours or days. Also, concentrate your searches on or just above the waves; seabirds do not migrate very high up.

In Britain and north-west Europe, excellent seabird migration can be enjoyed in April and May. But the best months, with the greatest number of birds and variety of species, are August and September.

## Escape Movements

Escape movements can be spectacular, albeit unpredictable. They can be seen at any place and at any time of day, although the early morning is usually productive. Most escape movements occur in winter, but they are also triggered by periods of midsummer drought. Some occur during blizzards, others in bright sunshine.

It is easy to overlook some escape movements because they involve familiar resident birds. Woodpigeon movements, for example, often go unnoticed by birdwatchers, and so do flocks of Starlings and gulls. Even crows may be caught up at times. You need to keep your eyes open for unusual numbers of familiar birds flying in one direction, not necessarily in large flocks, but also singly, one after the other.

Many birds escaping hard weather on the continent can be seen as they approach or fly along Britain's coastline, especially ducks and geese. Others may be seen anywhere inland. So with this highly spontaneous form of migration, the best advice is not to seek it but just to enjoy it when it happens.

## Moon Watching

If the prevailing conditions are favourable and the night is heavy with migration, it can be possible to see the silhouettes of birds as they pass across the Moon's face. You need a telescope – not an astronomical one, but the sort birdwatchers use – and the Moon should be as full as possible. Wait for a while and, before long, you should see some shapes flickering across the surface. You will not be able to identify the species concerned, and they will pass in an instant, but at least you will be able to see migration in action.

### Listening at Night

Night migrating birds often call to maintain contact if they are in a flock, although not all species do so. If you live in a quiet area, you can pick out the calls of quite a number of birds if you listen carefully. The easiest to hear are migratory thrushes, particularly Redwings and Fieldfares. On almost any starry night in October, especially just after dark, you are likely to hear the thin *seep* call of the former as it passes overhead.

*The Scarlet Rosefinch is the sort of bird that is easiest to see at a migration hotspot.*

## Places to See Migration

Some places are particularly good for watching migration, often because their localized topography concentrates bird movements. I have already mentioned spits of land that jut out from a coastline, acting as a first landfall and funnelling migrants together; excellent examples are Spurn Point in Yorkshire and Point Pelee in Ontario, Canada. Islands can be just as good. Heligoland in Germany, Falsterbö in Sweden, Texel in Holland, and Fair Isle in Scotland are all famous for their migrants, not to mention the Isles of Scilly off south-west England that in season attract hordes of rarity-watchers from all over Britain.

A few particularly favoured places have a permanent birdwatching presence in the form of a bird observatory. Essentially, this is a staffed ringing station (*see* pages 92–5) where the focus is on the scientific collection of facts. But observatories open their doors to everyone – ringer or non-ringer, obsessed birder or curious beginner – and they are often exciting places to visit, especially on a day of good migration.

**Always Welcome**

*Anyone can call in on an observatory to find out the local up-to-date news and to meet fellow birdwatchers prepared to share their experience and enthusiasm. If you've never been to an observatory, go!*

*The observatory on Fair Isle, between Orkney and Shetland.*

# 8
# GREAT JOURNEYS

*In this chapter we throw off any shackles of generalisation in order to describe some of the more interesting, celebrated or unusual bird migrations performed by European birds. These are a hand-picked group of my own choosing, and any number of other species could be selected and described, and probably deserve to be. Indeed we must remember and indeed celebrate the fact that no species' migration is quite the same as any other's.*

*White Storks leave Europe in late summer. They take either an eastern route via Turkey, the Middle East, and the Nile Valley to reach East Africa and even South Africa, or a Western route, via Gibraltar, to reach the North of Africa. They return to Europe in April.*

# ARCTIC TERN

No bird travels as far in a year as the Arctic Tern, the world record holder for the longest migration of all. It breeds in the far north, including well above the Arctic Circle, and then winters in the Antarctic, at nearly equivalent latitudes. It would actually be difficult to migrate any further unless you took a perverse west to east route instead. As far as figures are concerned, the Arctic Tern travels at least 35,000km in a year, clocking up 17,500km each way, and it is estimated that many individuals travel a great deal further than that, perhaps up to 50,000km annually.

Some Arctic Terns travel from the Arctic, with its Polar Bears…

If you ask why they do it, the answer is so simple that it makes you wonder why other birds don't do the same. For one thing, polar waters are often exceptionally rich fishing grounds, so the birds have little trouble finding enough food. Another perk of this near pole-to-pole migration is that Arctic Terns see more daylight each year than any other animal on Earth, giving them plenty of time to take advantage of this abundance. Their migration may be extreme, but it is highly expedient.

Some details of the Arctic Tern's travels are almost as impressive as the bald figures. For example, it appears that the populations breeding on the eastern side of North America do not just fly directly south, but actually cross the Atlantic as well, joining their Eurasian counterparts off the west coast of Africa. It is thought that these birds undertake their transoceanic flight at extreme high altitudes, using the prevailing westerly winds to power them. Another remarkable fact is that Arctic Terns do not rest when they arrive in Antarctic waters, but spread out west and east. Incredibly, it is thought that some actually circumnavigate the Antarctic continent – just a quick jaunt!

…to the Antarctic, with its Penguins.

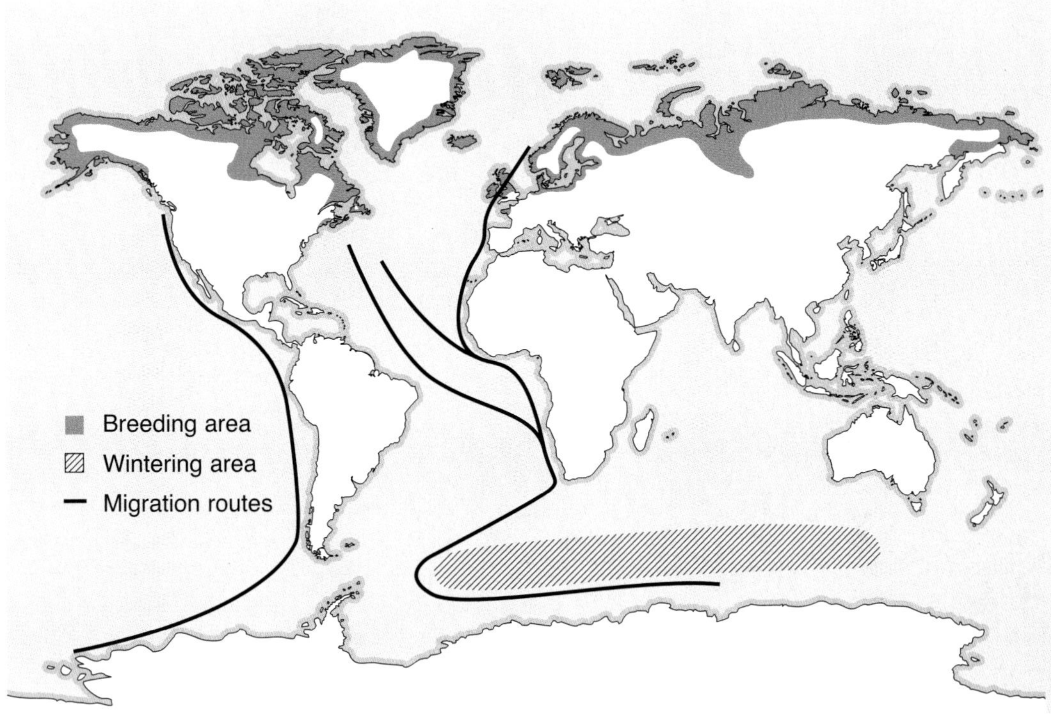


*Lines showing the approximate migration route of the Arctic Tern.*

On their return, most Arctic Terns appear to take the same route that they used to arrive, but there is also evidence that a few follow more of a loop migration (*see* page 14). There have been extraordinary ringing recoveries of birds in Russia's Ural Mountains, strongly suggesting that, far from following coastlines as they do in autumn, some Arctic Terns actually migrate overland at this season, no doubt at high altitude. Incidentally, Arctic Terns are common on inland reservoirs in spring in Britain; not an astonishing fact in itself, but at least proof that flying over land holds no fear for them.

> **A TOUGH OLD BIRD**
>
> *The oldest Arctic Tern known to date lived to the age of 33. With regular to and fro migration, it will have flown over a million kilometres in all during its lifetime. Even well-travelled human beings would be hard-pressed to rival this remarkable achievement.*

## SWALLOW

In Britain the Swallow is perhaps the most famous migrant of all, celebrated for its association with the coming of spring. This association is valid wherever the species goes, for it arrives in Europe and Asia from about March onwards and it also arrives in South Africa as the Southern Hemisphere summer approaches. It is worth noting that this same species – although under the different name of Barn Swallow – migrates between North and South America, where it is also viewed as a herald of fair weather.

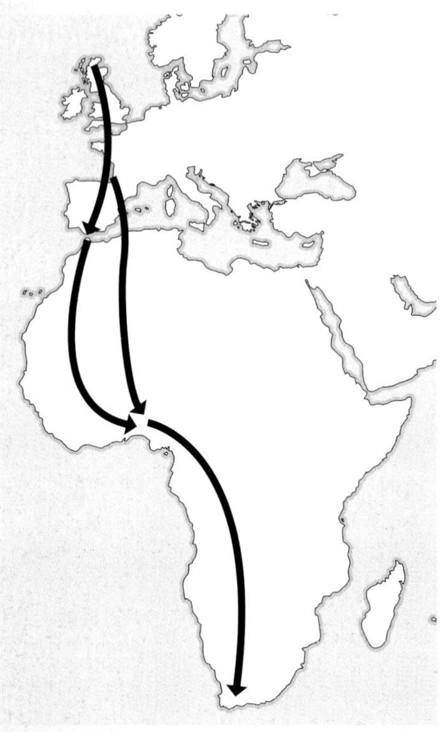

The migration route of British Swallows.

The Swallow is an unusual migrant because – in contrast to most small, insectivorous birds – it travels by day. In addition, like a marathon runner grabbing a drink without stopping, the Swallow can, at least in theory, take its sustenance *en route*. It is also rather easy to spot an individual in the act of migrating. Instead of sweeping to and fro over fields again and again, a migrating Swallow moves rapidly in its chosen direction, higher than usual and with an unfamiliar briskness.

Although Swallows undertake journeys from many parts of the world, the path they take from Britain southwards is particularly well known in its detail. It is recorded in the map opposite. Swallows start off slowly, moving south through this country in short bursts of a few tens of kilometres every few days and spending each night in communal roosts. By September, they are ready to cross the English Channel, and they do so on a broad front, not concentrating in the south-east of England. Having reached continental Europe, the Swallows then fly due south along the French coast in a narrow band only 100km or so wide, and find themselves at the foot of the western Pyrenees.

Interestingly, British Swallows now veer east, skirting the mountain chain until they reach the Mediterranean coast of Spain. Only at this point do they turn south again, either crossing the sea via the Balearic Islands or, more often, running down the east coast of Spain. They reach the southern tip of Spain and cross over into north-west Africa. The trail now goes a little cold (or hot, perhaps). It seems that most British Swallows are now in peak physical and

*Swallows – day time travellers.*

migratory health, and overfly the Sahara Desert in one hop. They abandon their usual custom of migrating by day on this leg, apparently preferring the darkness and perhaps the cool.

They turn up again in West Africa. There are ringing recoveries around the Gulf of Guinea, in Nigeria and in the Congo Basin near the coast. It is now almost November and the Swallows embark on the last stage of their journey, which to us is largely unknown. All we know for certain is that they turn up on their wintering grounds by December, having passed through a large swathe of tropical Africa with a comparatively sparse human population.

In contrast to their stay in Europe, which is a lengthy one, the Swallows remain in South Africa only until late February. Their return migration to Europe is decidedly brisk, with the first arrivals making landfall, even in Britain, during the month of March.

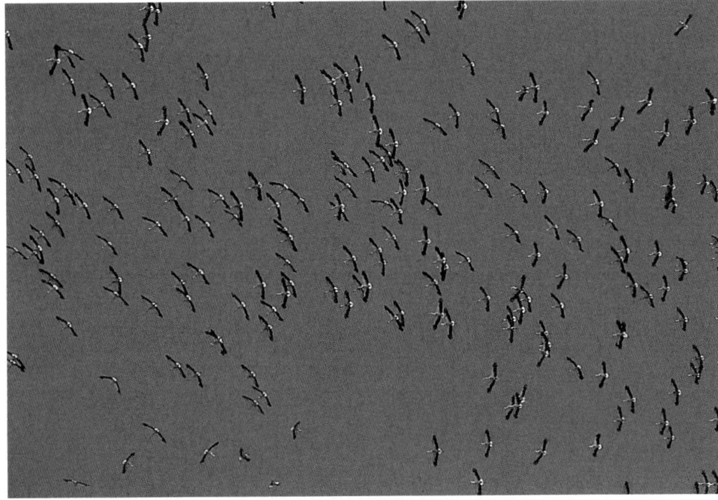

*White Storks are famous for migrating in huge, swirling flocks.*

## White Stork

The White Stork is a reliable migrant, in the sense that it turns up at its breeding sites at the same time each year after a sojourn to Africa. It is so reliable, in fact, that it has been entrusted in some cultures with the task of delivering babies. In central Europe the White Stork arrives in the first week of April, and it departs with the first whiff of autumn air, in August.

The stork's migration is a very long one – up to 11,000km each way – and yet, in theory at least, it requires little expenditure of energy. That's because storks have a very special form of travel: they ride thermals almost every step of the way. Thermals may be as much as 2km across, although they are usually much smaller.

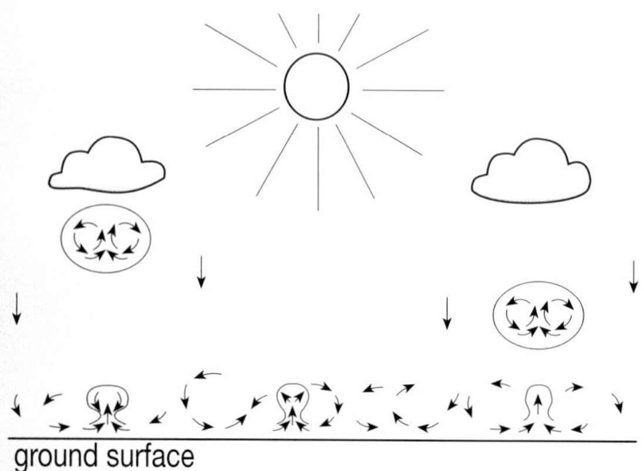


They form because hot air, being lighter than cool air, rises. Thermals come into being initially when the Sun's rays heat the ground; the adjacent air is heated, too, and begins lifting. This does not yet guarantee a thermal because the cooler air sinking down may cause too much turbulence, but if the rising stream is strong enough, it will break away from the rest of the layer of warmer air to rise as a bubble, often to thousands of metres above ground. The air in a thermal has a great deal of energy and more than enough strength to whisk a large bird very rapidly upwards.

*The formation of a thermal. The sun heats the air just above the ground, and it rises as a large bubble, producing strong swirling air currents that give the birds lift.*

There is a technique to travel by thermal. On its own a thermal updraft can only lift a bird up in a spiral rather than blowing it along in a certain direction. So the bird soars upwards as far as the thermal will take it, and then simply glides down towards another thermal situated in the correct migratory direction. It will of course lose height as it glides, but it will soon be carried upwards again by the next rising spiral. And so the bird travels along its route, thermal by thermal, with effortless ease.

*The 'veins' shown on the map indicate the preferred migration routes of White Storks.*

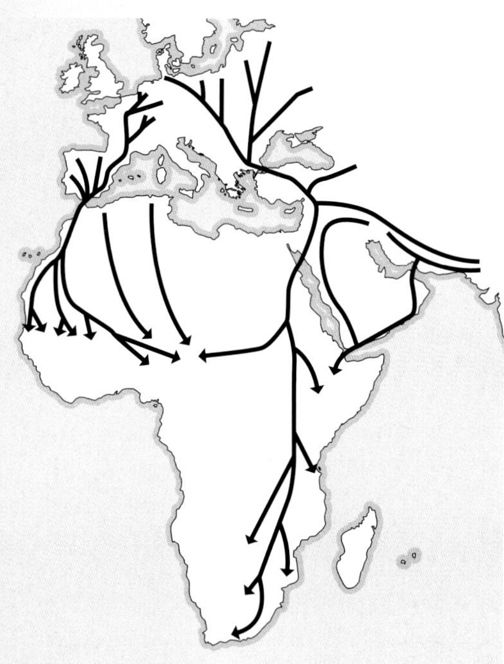

The only problem with thermal travel is that it is limited to places where thermals develop and to the times when thermals develop. One of the White Stork's problems is that thermals do not form over water, so its migratory route has to direct it to a narrow sea crossing, such as the ones at southernmost Spain and the Bosphorus (*see* map). Storks also have to migrate by day and since it takes quite some time for thermals to form – not until mid-morning even on the hottest days – their travelling time is limited. They could, of course, revert to flapping and gliding flight

## Mass Movement

*Migrating storks are a spectacular sight, not only because of their spiralling from thermal to thermal, but also because of the size of their flocks. Up to 11,000 storks have been seen in a single movement.*

outside these times if they needed to, but it has been calculated that this takes over ten times as much energy.

## Northern Wheatear

The Northern Wheatear has one of the longest migratory journeys of all birds, which is special enough. But its route also admirably demonstrates that a migratory path is not always expedient. Sometimes it is thoroughly eccentric, and very much a product of tradition and history.

There is nothing particularly unusual about the journeys of Northern Wheatears breeding in Europe. They simply fly to Africa and back, wintering in a broad belt south of the Sahara. However, the Northern Wheatear is a very widespread species: it breeds eastwards all across the north of Asia, spilling across into Alaska. To the west of Europe it also breeds in Greenland and northern Canada.

One might expect the outlying populations, especially the North American ones, to be 'sensible' and migrate south to Central or South America, cutting thousands of kilometres off their journey. At the same time, you might expect the north Asian populations to go to India. But they don't. Unbelievably, they all go to Africa which, in the case of the Greenland Wheatear (a subspecies

*Regardless of where they breed, all Northern Wheatears winter in Africa.*

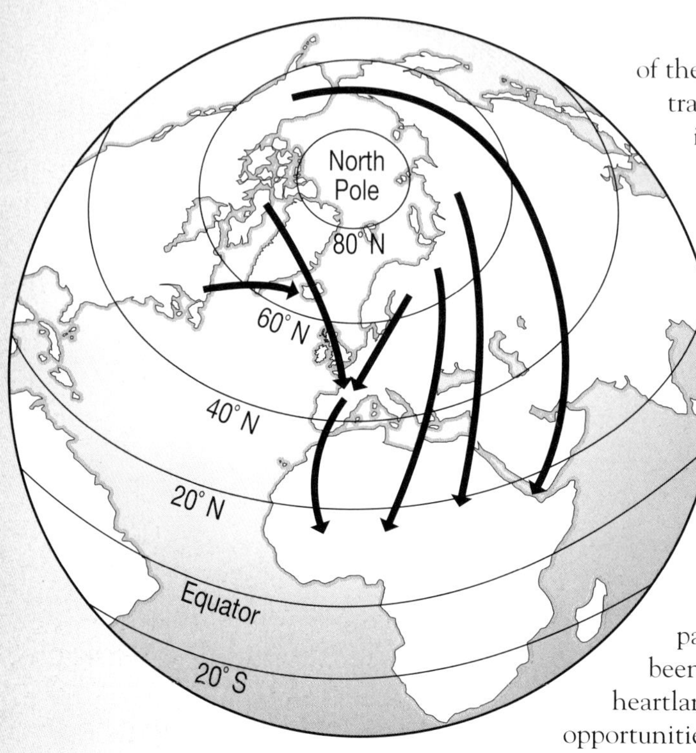

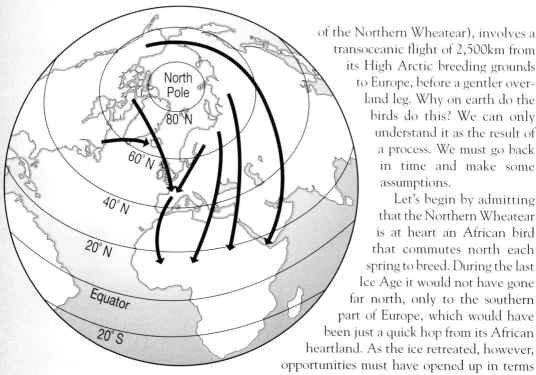

Migration routes of the Wheatear, showing that even the birds of Alaska and Greenland make their way to Africa.

of the Northern Wheatear), involves a transoceanic flight of 2,500km from its High Arctic breeding grounds to Europe, before a gentler overland leg. Why on earth do the birds do this? We can only understand it as the result of a process. We must go back in time and make some assumptions.

Let's begin by admitting that the Northern Wheatear is at heart an African bird that commutes north each spring to breed. During the last Ice Age it would not have gone far north, only to the southern part of Europe, which would have been just a quick hop from its African heartland. As the ice retreated, however, opportunities must have opened up in terms of reduced competition and increased daylight, and the population edged north.

If north, why not east and west? We can imagine that the current range of the Northern Wheatear came about step by step, inching west or east each year, with the more adventurous birds making the advance over the eastern Atlantic or the Bering Straits. And if the breeding range expanded slowly enough, the wheatears would not have 'noticed' that their return flight was getting longer and longer.

Greenland Wheatears are slightly larger than other Northern Wheatears.

## Great Shearwater

Although the Arctic Tern is credited with longest migration of any species of bird, several other seabirds come close, one of which is the Great Shearwater. In a given year each individual of this species travels at least 30,000km, journeying from high southern latitudes to high northern latitudes and back again.

The direction taken by the Great Shearwater is shown in the diagram, and you'll see that it is a loop migration *par excellence*. The birds breed on islands in the south Atlantic, notably at

Tristan da Cunha, between October and April each year, and then travel north along the eastern seaboard of both South and North America. They progress rapidly through tropical regions and arrive off Newfoundland late in the northern spring. They then move slightly east to the seas off southwest Greenland in July and August, where they moult and languish, flightless, in the abundantly nutrient-rich waters there. Refreshed and re-feathered they then move south again, nearing the seaboard of the western Atlantic, where they may be seen off the western coasts of Britain and Ireland, as well as the Bay of Biscay, Portugal and the Azores. It is then a quick dash non-stop across the Atlantic Ocean to reach their breeding islands again by September.

*Great Shearwaters.*

This phenomenal annual journey is aided by the Great Shearwater's use of the ocean's energy. One very efficient type of flight locomotion that is uses is known as **slope soaring**. Waves travelling through the water force air up over them, and if the shearwater can remain in these updrafts by moving along and in front of the waves, it can simply ride along for hours at a time without having to flap its wings. By slope-soaring, a shearwater can travel a long way very fast without expending much energy at all. No doubt about it: this is migration in First Class!

In higher winds the Great Shearwater may take a more spectacular ride. In a manoeuvre known as **dynamic soaring**, the shearwater slides down to meet a wave's updraft from some height. As it accelerates down, as if on a child's slide, the shearwater's momentum builds up so that, when it hits the updraft, it is

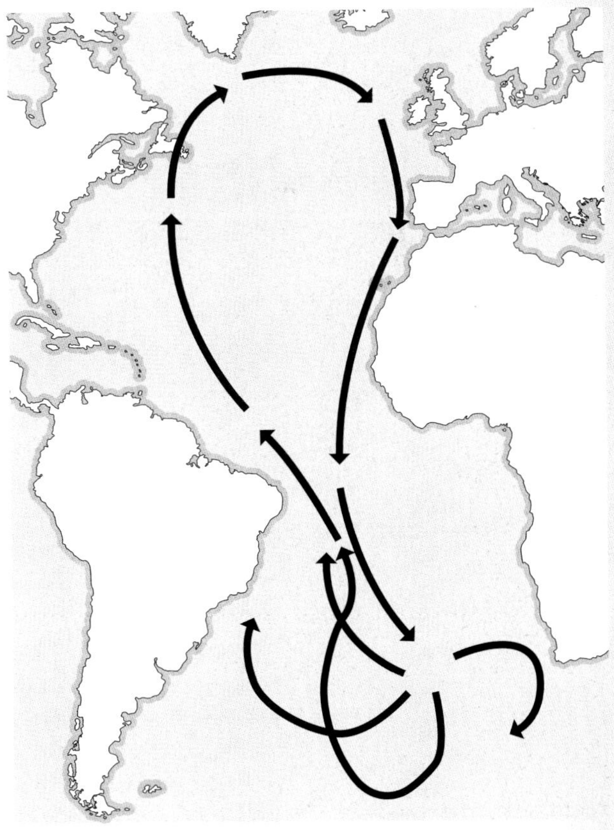

*The migration route of the Great Shearwater, showing its gigantic loop.*

swept up high again. In common with slope soaring, all the energy required for this is donated by the elements, and the bird can travel great distances with scarcely a wing beat.

## Eleonora's Falcon

The lives of few birds are as wrapped up in migration as that of Eleonora's Falcon, a rather rare species that breeds in colonies on offshore islands in the Mediterranean and eastern Atlantic. It is a major predator on migrating birds and it undertakes a highly bizarre and noteworthy journey of its own: it is the only European bird that winters on the large Indian Ocean island of Madagascar.

Eleonora's Falcons arrive in their breeding areas from late April, and for the next few months feed mainly on insects. They are streamlined and aerobatic hunters, well able to handle any evasion manoeuvres that might be attempted by a dragonfly, moth or plump beetle. The falcons continue feasting on insects until mid-July when, much later than most other local species, they finally lay eggs. These hatch by the end of August and – lo and behold – the southward migration of birds is now in full swing, so the falcons switch to this abundant food supply to nourish their hungry chicks. In other words, Eleonora's Falcon times its breeding season specifically to take advantage of the passing of vast numbers of small migrants on their way to Africa.

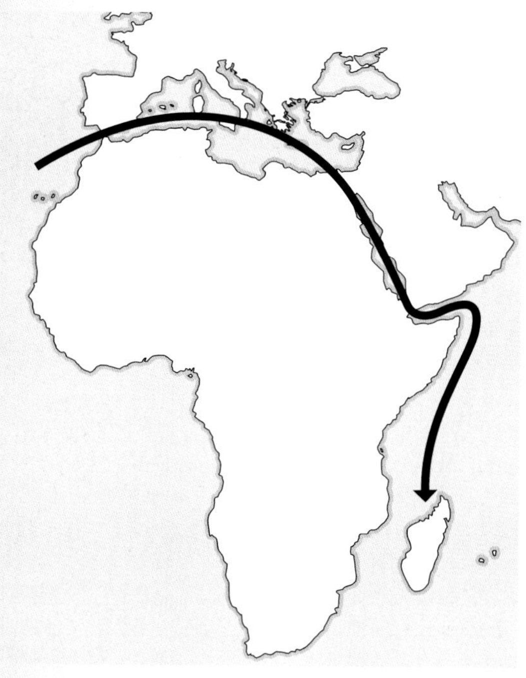

*The remarkable migration route of the Eleonora's Falcon.*

By living on offshore islands, the falcons can concentrate on catching birds as they migrate over the sea. Several colonies of falcons are found at sites that mark the end of a long sea crossing, thus allowing the interception of birds that are especially tired. With their swift, agile flight, Eleonora's Falcons are formidable predators on their own, but the fact that they live socially makes a colony a daunting prospect for any small bird that might need to pass by.

Members of a colony may be seen spreading out over the sea and hovering or holding their position in a line, a bit like border guards placed along a frontier, but also at different heights. They might also adopt a different strategy, patrolling widely over a large area, catching and killing what they

come upon by chance. The falcons have a habit of harrying intercepted birds towards the waves, slowly undermining their attempts to escape as exhaustion overtakes their unfortunate prey. It is estimated that, on any given autumn day, the Eleonora's Falcons of the Mediterranean and eastern Atlantic take 30,000 birds, or more than two million in the course of a single season.

*Eleonora's Falcons wait in line, each at its post, intending to intercept tired migrants.*

Having taken their toll of migrants, the falcons become migrants themselves by the end of October. It appears that all or most of them move east along the Mediterranean coast and then south-east through the Red Sea, before following the Indian Ocean south to Madagascar and resuming their insect diet. Although a few of the falcons are found in East Africa, the question is: why do most go to Madagascar? The answer is a little speculative, but it would appear that they do so to avoid competition for food. Several other insect-eating falcons winter in sub-Saharan Africa, including Hobbies and Red-footed Falcons; it would appear that Madagascar is the only place at similar latitude where there is still a vacant niche to be exploited.

# MIGRATORY TABLE FOR BRITISH BIRDS

• There are six crosses allocated to each species in the left hand columns. The numbers of crosses in each column are a reflection of the proportion of the population that are relevant to that column. For example, if a bird is exclusively a summer visitor it will have six crosses in that section. If some of the population are also found here in the winter, some of the crosses will be donated to the resident section, and so on.

| | *Resident* | *Summer Visitor* | *Winter Visitor* | *Mainly Passage* | *Migrates within Britain* | *Summer Season* | | | |
|---|---|---|---|---|---|---|---|---|---|
| | | | | | | *Initial arrival* | *Main arrival* | *Departure* | *Winter destination* |
| Red-throated Diver | XXX | | XXX | | Yes | | | | |
| Black-throated Diver | XX | | XXXX | | | | | | |
| Great Northern Diver | | | XXXXXX | | | | | | |
| Little Grebe | XXXX | | XX | | Yes | | | | |
| Great Crested Grebe | XXXXX | | X | | Yes | | | | |
| Red-necked Grebe | X | | XXXXX | | | | | | |
| Slavonian Grebe | XXX | | XXX | | Yes | | | | |
| Black-necked Grebe | XX | | XXXX | | Perhaps | | | | |
| Northern Fulmar | XXXX | XX | | | | 3 | 4 and 5 | 7 to 12 | Atlantic and North Sea |
| Manx Shearwater | | XXXXXX | | | | late 2 | 3 and 4 | 7 to 9 | South Atlantic |
| European Storm-petrel | | XXXXXX | | | | late 4 | 5 | 9 to 10 | South African waters |
| Leach's Storm-petrel | | XXXXXX | | | | late 4 | 5 | 9 to 10 | Oceanic Equatorial waters |
| Northern Gannet | XXX | XXX | | | | 1 | 2 to 3 | 10 | African coast to Senegal |
| Great Cormorant | XXXXXX | | | | Yes | | | | |
| Shag | XXXXXX | | | | Yes, not far | | | | |
| Bittern | XXX | | XXX | | Yes | | | | |
| Little Egret | XXX | | XXX | | | | | | |
| Grey Heron | XXXX | | XX | | | | | | |
| Mute Swan | XXXXXX | | | | | | | | |
| Bewick's Swan | | | XXXXXX | | | | | | |
| Whooper Swan | | | XXXXXX | | | | | | |
| Bean Goose | | | XXXXXX | | | | | | |
| Pink-footed Goose | | | XXXXXX | | | | | | |
| White-fronted Goose | | | XXXXXX | | | | | | |
| Greylag Goose | XXX | | XXX | | | | | | |
| Canada Goose | XXXXXX | | | | | | | | |
| Barnacle Goose | | | XXXXXX | | | | | | |
| Brent Goose | | | XXXXXX | | | | | | |
| Egyptian Goose | XXXXXX | | | | | | | | |
| Common Shelduck | XX | XXXX | | | Some | 9 of previous year | 10 to 12 of previous year; 1 | 7 and 8 | Waddenzee of N Germany |
| Mandarin Duck | XXXXXX | | | | | | | | |
| Eurasian Wigeon | X | | XXXXX | | Perhaps | | | | |
| Gadwall | XX | X | XXX | | | ? | ? | 7 onwards | France, Low Countries, Mediterranean |
| Teal | XX | | XXXX | | | | | | |
| Mallard | XXXX | | XX | | | | | | |
| Pintail | X | | XXXXX | | Few | | | | |
| Garganey | | XXXXXX | | | | 3 | 4 | 7 to 10 | Tropical Africa |
| Shoveler | XX | X | XXX | | | 3 | 4 | 7 to 11 | France, Spain Mediterranean |
| Common Pochard | XX | X | XXX | | Yes | 2 | 4 | 6 to 11 | |
| Tufted Duck | XXX | | XXX | | Yes | | | | |
| Greater Scaup | | | XXXXXX | | | | | | |
| Eider | XXXX | | XX | | Yes | | | | |
| Long-tailed Duck | | | XXXXXX | | | | | | |
| Common Scoter | X | | XXXXX | | | | | | |
| Velvet Scoter | | | XXXXXX | | | | | | |
| Common Goldeneye | XX | | XXXX | | | | | | |
| Smew | | | XXXXXX | | | | | | |
| Red-breasted Merganser | XXX | | XXX | | Yes | | | | |
| Goosander | XX | XXXX | | | | | | | |
| Ruddy Duck | XXXXX | X | | | | 3? | 4? | 7 to 10? | Continental Europe |
| Honey Buzzard | | XXXXXX | | | | mid 5 | late 5 | mid 8 to 9 | Tropical Africa |
| Red Kite | XXXXXX | | | | | | | | |
| White-tailed Eagle | XXXXXX | | | | | | | | |
| Marsh Harrier | X | XXXXX | | | | 3 | early 4 | prob 9 to 10 | North-west or West Africa |

• The numbers in the table relate to months; 1 for January, 2 for February, etc.
• If an element of a population 'Migrates within Britain' it is still deemed 'resident' since it does not leave this country.
• If a bird has an entry in the 'Mainly Passage' column it means that either a bird is entirely a passage migrant, or that a very high proportion of birds seen in this country are only passing through, swamping those that are arriving or leaving here.

| *Winter Season* | | | *Hard-weather Movements* | *Moult Migration* | *Altitudinal Migration* | *Irruptions into Britain?* | *Notes* |
|---|---|---|---|---|---|---|---|
| *Main Arrival* | *Main Departure* | *Breeding destination* | | | | | |
| late 8 | late 2 to 5 | Northern Europe and Iceland | | | | | |
| 9 and 10 | 4 | Northern Europe | | | | | |
| 8 | by early May | Iceland | | | | | |
| ? | ? | Northern Europe | | | | | |
| 8 to 10 | mid 2 onwards | North-west Europe | | | | | |
| 10 | 3 | Northern Europe | | | | | |
| 10 and 11 | 3 | Iceland, Scandinavia | | | | | |
| 10 and 11 | 3 | Northern Europe | | | | | |
| | | | | | | | |
| | | | | | | | |
| | | | | | | | |
| | | | Wrecks in autumn | | | | |
| | | | | | | | |
| | | | Inland in strong winds; to coast when freezing inland. | | | | |
| | | | | | | | |
| 9 to 12 | 3? | N and E Europe | Yes | | | | |
| 9 and 10 | 3 | Continental Europe, especially France | | | | | Juveniles highly dispersive |
| 9 to 10 | 3 | North-west Europe | | | | | |
| | | | Some influxes in winter | Yes | | | |
| 10 to 12 | 2 to 3 | Russian tundra | Yes | | | | |
| mid 10 | 3 to mid 4 | Iceland | | | | | |
| late 9 to early 10 | 3 | Scandinavia | | | | | |
| 10 | 4 | Greenland and Iceland | | | | | |
| 10 to 12 | 3 to 4 | Greenland and Russia | | | | | |
| late 9 to early 11 | mid 3 to late 4 | Iceland | | | | | |
| | | | A few leave Britain | Yes | | | |
| 10 | late 3 to 4 | Greenland, Spitzbergen, Siberia | | | | | |
| 10 | mainly 3 | Greenland, Russia | | | | | |
| | | | | | | | |
| Yes | | | | Yes | | | Mainly a moult migration |
| | | | | | | | |
| 10 and 11 | 2nd half 3 to 4 | Northern Europe and Russia | Yes | Yes | | | |
| 9 to 11 | 3 and 4 | Northern and Eastern Europe | Yes | Yes | | | |
| 7 to 11 | late 2 to 5 | Iceland, Scandinavia to northwest Siberia | Pronounced | Yes | | | |
| | | North and East Europe | | Yes | | | |
| 9 to 10 | 3 to 4 | Northern Europe | Yes | Yes | | | |
| | | | | Yes | | | |
| 8 to 11 | 3 to 4 | Northern Europe | Yes | | | | |
| 6 to 11 | 2 to 4 | Baltic countries and Russia | Yes | Yes | | | |
| 7 to 10 | late 2 to 5 | Russia, Scandinavia, Iceland | Yes | Yes | | | |
| late 10 | 2 to 3 | Northern Europe | | | | | |
| 9 and 10 | 3 and 4 | Baltic and Netherlands | Yes | | | | |
| mid 10 to 12 | late 2 to 5 | Northern Europe | | | | | |
| mid 8 to 12 | late 2 to early 4 | Scandinavia and Iceland | Yes | | | | |
| mid 9 to 11 | mid 3 to 5 | Scandinavia and Russia | | | | | |
| 9 and 10 | 3 to 5 | Scandinavia | | | | | |
| 10 to 12 | late 2 to 4 | Mainly European Russia | | | | | |
| 10 to 12 | late 2 to 4 | Northern and Central Europe, Iceland | | Yes | | | |
| 10 to 12 | late 2 to 4 | Scandinavia | | Yes | | | |
| | | | | | | | Becoming increasingly migratory |
| | | | | | | | |
| | | | | | | | Young disperse widely |
| | | | | | | | |
| | | | | | | | |

| | Resident | Summer Visitor | Winter Visitor | Mainly Passage | Migrates within Britain | Summer Season | | | |
|---|---|---|---|---|---|---|---|---|---|
| | | | | | | Initial arrival | Main arrival | Departure | Winter destination |
| Hen Harrier | XXX | | XXX | | Yes | | | | |
| Montagu's Harrier | | XXXXXX | | | | 4 | 5 | 7 to 9 | Tropical and Southern Africa |
| Goshawk | XXXXXX | | | | | | | | |
| Sparrowhawk | XXXXXX | | | | | | | | |
| Common Buzzard | XXXXX | | X | | | | | | |
| Rough-legged Buzzard | | | XXXXXX | | | | | | |
| Kestrel | XXXXXX | | | | | | | | |
| Merlin | XXX | | XXX | | Yes | | | | |
| Hobby | | XXXXXX | | | | Late 4 | 5 | 8 to 10 | Southern Africa |
| Peregrine | XXXXXX | | | | | | | | |
| Red Grouse | XXXXXX | | | | | | | | |
| Ptarmigan | XXXXXX | | | | | | | | |
| Black Grouse | XXXXXX | | | | | | | | |
| Capercaillie | XXXXXX | | | | | | | | |
| Red-legged Partridge | XXXXXX | | | | | | | | |
| Grey Partridge | XXXXXX | | | | | | | | |
| Quail | | XXXXXX | | | | late 4 | 5 | ? | Central Africa |
| Pheasant | XXXXXX | | | | | | | | |
| Golden Pheasant | XXXXXX | | | | | | | | |
| Lady Amherst's Pheasant | XXXXXX | | | | | | | | |
| Water Rail | XXXX | | XX | | | | | | |
| Spotted Crake | | XXXXXX | | | | 3 | 4 and 5 | 8 to 11 | Central and East Africa |
| Corncrake | | XXXXXX | | | | late 4 | 5 | late 7 to 9 | South-east Africa |
| Moorhen | XXXX | | XX | | | | | | |
| Coot | XXX | | XXX | | | | | | |
| Oystercatcher | | | | | | | | | |
| Avocet | | XXX | XXX | | Yes | mid 3 | late 3 to mid 4 | 8 and 9 | Spain and Portugal |
| Stone Curlew | | XXXXXX | | | | mid 3 | rest of 3 | 10 and 11 | Mediterranean |
| Little Ringed Plover | | XXXXXX | | | | early 3 | rest of 3 | 7 to 9 | West Africa? |
| Ringed Plover | XXX | | XXX | | | | | | |
| Dotterel | | XXXXXX | | | | mid 4 | late 4 to 5 | 7 and 8 | Spain and North Africa |
| Golden Plover | XXX | | XXX | | Yes | | | | |
| Grey Plover | | | XXXXXX | | | | | | |
| Lapwing | XX | XX | XX | | | late 2 | 3 and 4 | 7 to 9 | Ireland, France, Spain |
| Knot | | | XXXXXX | | | | | | |
| Sanderling | | | XXXXXX | | | | | | |
| Little Stint | | | | XXXXXX | | | | | |
| Curlew Sandpiper | | | | XXXXXX | | | | | |
| Purple Sandpiper | | | XXXXXX | | | | | | |
| Dunlin | | XXX | XXX | | | early 4 | late 4 to 5 | late 6 to 9 | Mainly West Africa? |
| Ruff | | | XX | XXXX | | | | | |
| Jack Snipe | | | XXXX | XX | | | | | |
| Snipe | XXX | | XXX | | Yes | | | | |
| Woodcock | XXX | X | XX | | | | 4 | 9 and 10 | Ireland and France |
| Black-tailed Godwit | XX | | XXXX | | | | | | |
| Bar-tailed Godwit | | | XXXXXX | | | | | | |
| Whimbrel | | XX | | XXXX | | mid 4 | 5 | 7 to 9 | West to South Africa |
| Curlew | XXX | | XXX | | Yes | | | | |
| Spotted Redshank | | | X | XXXXX | | | | | |
| Redshank | XXX | | XXX | | | | | | |
| Greenshank | X | X | X | XXX | Yes | mid 3 | late 3 | mid 6 to 8 | South-west Europe, North Africa? |
| Green Sandpiper | | | XX | XXXX | | | | | |
| Wood Sandpiper | | | | XXXXXX | | | | | |
| Common Sandpiper | | XXXX | | XX | | 3 | 4 | 7 to 9 | Central and Southern Africa |
| Turnstone | | | XXXXXX | | | | | | |
| Red-necked Phalarope | | XXXXX | | X | | early 5 | mid 5 to early 6 | 7 and 8 | Arabian Sea |
| Grey Phalarope | | | | XXXXXX | | | | | |
| Arctic Skua | | XXX | | XXX | | mid 4 | late 4 | 8 and 9 | Not known |
| Long-tailed Skua | | | | XXXXXX | | | | | |
| Pomarine Skua | | | | XXXXXX | | | | | |
| Great Skua | | XXX | | XXX | | late 3 | 4 to 5 | 8 to 10 | Atlantic seaboard to North Africa |
| Mediterranean Gull | X | | XXXXX | | | | | | |
| Little Gull | | | | XXXXXX | | | | | |
| Black-headed Gull | XXX | | XXX | | Yes | | | | |
| Common Gull | XX | | XXXX | | Yes | | | | |
| Lesser Black-backed Gull | X | XXX | XX | | Yes | late 1 | mid 2 to 4 | late 7 to 10 | Iberia and North Africa |
| Herring Gull | XXX | | XXX | | Yes | | | | |
| Yellow-legged Gull | | | XX | XXXX | | | | | |
| Glaucous Gull | | | XXXXXX | | | | | | |
| Iceland Gull | | | XXXXXX | | | | | | |

| Winter Season | | | Hard-weather Movements | Moult Migration | Altitudinal Migration | Irruptions into Britain? | Notes |
|---|---|---|---|---|---|---|---|
| Main Arrival | Main Departure | Breeding destination | | | | | |
| 8 to 10 | 2 and 3 | Northern Europe | | | | | |
| | | | | | | | |
| | | | | | | | |
| | | | | | | | |
| 10 | 3 | Unknown | | | | | |
| 10 | 2 | Scandinavia and Russia | | | | | |
| | | | | | | | |
| 9 and 10 | 3 | Iceland | | | | | |
| | | | | | | | |
| | | | | | Yes | | |
| | | | | | Yes | | |
| | | | | | Yes | | |
| | | | | | | | |
| | | | | | | | |
| | | | | | | | |
| | | | | | | | |
| | | | | | | | May be second arrival in July |
| | | | | | | | |
| | | | | | | | |
| | | | | | | | |
| 10 and 11 | 3 and 4 | North and East Europe | | | | | |
| | | | | | | | |
| | | | | | | | |
| 8 to 11 | ? | North-west Europe | | | | | |
| 7 to 9 | 1 to 4 | North-west Europe | Yes | Yes | | | |
| 8 to 10 | 3 and 4 | Iceland, North-west Europe | | | | | |
| | | North-west Europe | | | | | |
| | | | | | | | |
| | | | | | | | |
| 8 to 9 | 4 and 5 | Northern Europe | | | | | |
| | | | | | | | |
| 8 to 10 | 4 to 5 | Iceland and Scandinavia | Yes, to south-west | Yes | | | |
| end 7 to 10 | 4 and 5 | Arctic Russia | | | | | |
| late 9 to early 11 | 2 to 4 | Central Europe, Russia | Yes, to south-west | | | | |
| 8 onwards | 5 | Arctic Canada | | | | | |
| mid 7 to 10 | 3 to mid 5 | Greenland, Siberia | | | | | |
| | | | | | | | late 7 to 10 |
| | | | | | | | Mainly 8 and 9 |
| late 8 to 10 | 2 to 5 | Northern Europe (esp Norway) and Greenland | | | | | |
| late 6 to 10 | 3 to 5 | Scandinavia and Russia | | | | | |
| 7 to 10 | 3 and 4 | Scandinavia | | | | | |
| 8 to 11 | 2 to 4 | Scandinavia and probably Russia | | | | | |
| 9 to 11 | 3 to 5 | Iceland and Scandinavia | | | Yes | | |
| 10 to 11 | to mid 4 | Finland, Russia, Sweden, Norway | | | | | |
| 6 to 9 | 3 to 5 | Iceland | | | | | |
| 7 to 10 | 3 and 4 | Arctic | | | | | |
| | | | | | | | |
| late 6 to 10 | 3 to 5 | Northern Europe | | | | | |
| mid 8 to mid 9 | late 3 to early 5 | Scandinavia | | | | | |
| 7 to 10 | 3 and 4 | Mainly Iceland, some Scandinavia | | | | | |
| 7 to 10 | 4 to 5 | North-east Europe | | | | | |
| mid 6 to 10 | mid 4 to 5 | Northern Europe | | | | | |
| | | | | | | | Mainly 5 and mid 7 to 9 |
| | | | | | | | |
| 8 to 10 | 5 | Northern Europe, Greenland, NE Canada | | | | | |
| | | | | | | | |
| | | | Wrecks | | | | Mainly 9 to 11 |
| | | | | | | | |
| | | | | | | | Mainly 5 and 8 to 9 |
| | | | | | | | 5 and 9 to 11 |
| | | | | | | | |
| 7 to 10 | 2 to 4 | Central Europe and Baltic | | | | | |
| | | | | | | | Mainly 3 to 5 and 8 to 10 |
| [illegible] to 12 | 2 and 3 | Northern and Eastern Europe | | | | | |
| 8 to 12 | 3 and 4 | Northern Europe (Scandinavia, Russia) | | | | | |
| [illegible] to 11 | 1 to mid 4 | Northern Europe, esp Scandinavia | | | | | |
| [illegible] and 10 | 2 | Northern Europe | | | | | |
| [illegible] to 11 | ? | Near continent? | | | | | |
| [illegible]1 | 3 | Arctic | | | | | |
| [illegible]0 to 12 | 3 | Greenland and Arctic Canada | | | | | |

| | Resident | Summer Visitor | Winter Visitor | Mainly Passage | Migrates within Britain | Summer Season | | | |
|---|---|---|---|---|---|---|---|---|---|
| | | | | | | Initial arrival | Main arrival | Departure | Winter destination |
| Great Black-backed Gull | XXX | | XXX | | | | | | |
| Sabine's Gull | | | | XXXXXX | | | | | |
| Kittiwake | XXX | XXX | | | Yes | 2 | 3 and 4 | 7 and 8 | Atlantic (including west) |
| Sandwich Tern | | XXXXXX | | | | late 3 | 4 and 5 | 7 to 10 | African coast |
| Roseate Tern | | XXXXXX | | | | early 5 | mid 5 | 8 to 9 | West African Coast |
| Common Tern | | XXXXXX | | | | early 4 | mid 4 to late 5 | 8 to 10 | West African Coast |
| Arctic Tern | | XXXXXX | | | | 5 | 5 to 6 | 7 to early 10 | Antarctic |
| Little Tern | | XXXXXX | | | | 4 | 5 | 7 to early 10 | West African Coast |
| Black Tern | | | | XXXXXX | | | | | |
| Guillemot | XXX | XXX | | | | 3 | 4 | early 8 to 9 | North Sea, Atlantic |
| Razorbill | X | XXXX | X | | | 3 | 4 | 7 to 10 | North Sea, Coast of France and Iberia, North Africa |
| Black Guillemot | XXXXXX | | | | Yes, not far | | | | |
| Puffin | | XXXXXX | | | | early 3 | late 3 and 4 | 8 | North Sea and Atlantic |
| Little Auk | | | XXXXXX | | | | | | |
| Feral Pigeon | XXXXXX | | | | | | | | |
| Stock Dove | XXXXXX | | | | | | | | |
| Woodpigeon | XXXXXX | | | | | | | | |
| Collared Dove | | | | | | | | | |
| Turtle Dove | | XXXXXX | | | | mid 4 | 5 | 7 to 9 | Central Africa |
| Cuckoo | | XXXXXX | | | | mid 4 | late 4 and 5 | 8 to 9 | Central and Southern Africa |
| Ring-necked Parakeet | XXXXXX | | | | | | | | |
| Barn Owl | XXXXXX | | | | | | | | |
| Little Owl | XXXXXX | | | | | | | | |
| Tawny Owl | XXXXXX | | | | | | | | |
| Long-eared Owl | XXXX | | XX | | | | | | |
| Short-eared Owl | | XXX | XXX | Yes | | mid 4 | late 4 to 5 | 7 to 10 | Iberia and Southern Europe |
| Nightjar | | XXXXXX | | | | late 4 | 5 | 7 to 9 | Central and Southern Africa |
| Swift | | XXXXXX | | | | late 4 | 5 | 8 | Southern Africa |
| Kingfisher | XXXXXX | | | | | | | | |
| Hoopoe | | | | XXXXXX | | | | | |
| Wryneck | | | | XXXXXX | | | | | |
| Green Woodpecker | XXXXXX | | | | | | | | |
| Great Spotted Woodpecker | XXXXXX | | | | | | | | |
| Lesser Spotted Woodpecker | XXXXXX | | | | | | | | |
| Woodlark | XXXXX | X | | | Some | late 1 | 3 | 9 to 10 | |
| Skylark | XXXXX | | X | | | | | | |
| Shore Lark | | | | XXXXXX | | | | | |
| Sand Martin | | XXXXXX | | | | mid 3 | 4 | 8 | Sub-Saharan Africa |
| Swallow | | XXXXXX | | | | late 3 | 4 | 9 to 10 | South Africa |
| House Martin | | XXXXXX | | | | 4 | 5 | 9 to 10 | Africa (unknown where) |
| Tree Pipit | | XXXXXX | | | | late 3 | late 4 to early 5 | late 8 to 10 | Tropical Africa |
| Meadow Pipit | XXX | | XXX | | Yes | | | | |
| Rock Pipit | XXXX | | XX | | | | | | |
| Water Pipit | | | XXXXXX | | | | | | |
| Yellow Wagtail | | XXXXXX | | | | late 3 | 4 to 5 | 7 to 8 | West Africa |
| Grey Wagtail | XXXXXX | | | | Yes | | | | |
| Pied Wagtail | XXXX | XX | | | Yes | 3 | 4 | late 9 to 10 | Iberia and North Africa |
| Waxwing | | | XXXXXX | | | | | | |
| Dipper | XXXXXX | | | | | | | | |
| Wren | XXXXXX | | | | | | | | |
| Dunnock | XXXXXX | | | | | | | | |
| Robin | XXXXX | X | | | Yes | 3 | 4 | 10? | South-west Europe? |
| Nightingale | | XXXXXX | | | | mid 4 | late 4 to early 5 | 7 to early 9 | Africa |
| Black Redstart | XX | | X | XXX | Yes | | | | |
| Redstart | | XXXXXX | | | | early 4 | mid 4 to mid 5 | late 8 to 10 | Central Africa |
| Whinchat | | XXXXXX | | | | mid 4 | late 4 to 5 | late 7 to end 10 | Tropical Africa |
| Stonechat | XXX | XXX | | | Yes | late 2 | 3 | 9 | Southern Iberia |
| Northern Wheatear | | XXXXXX | | | | early 3 | late 3 to 4 | mid 7 to late 10 | Sub-Saharan Africa |
| Ring Ouzel | | XXXXXX | | | | mid 3 | late 3 | late 8 to 10 | Mediterranean Basin |
| Blackbird | XXXXX | | X | | | | | | |
| Fieldfare | | | XXXXXX | | | | | | |
| Song Thrush | XXXXX | | X | | | | | | |
| Redwing | | | XXXXXX | | | | | | |

| Winter Season | | | Hard-weather Movements | Moult Migration | Altitudinal Migration | Irruptions into Britain? | Notes |
|---|---|---|---|---|---|---|---|
| *Main Arrival* | *Main Departure* | *Breeding destination* | | | | | |
| 7 | 2 | Scandinavia | | | | | |
| | | | | | | | Mainly 9 and 10 |
| | | | | | | | |
| | | | | | | | |
| | | | | | | | |
| | | | | | | | |
| | | | | | | | |
| | | | | | | | |
| | | | | | | | Mainly 7 to 9 and 5 |
| | | | | | | | |
| 9 to 11 | ? | Northern Europe | | | | | |
| | | | | | | | |
| | | | Wrecks | | | | |
| 10 to 12 | 2 | Probably Spitzbergen and Arctic Russia | Wrecks | | | | |
| | | | | | | | |
| | | | | | | | |
| | | | | | | | |
| | | | | | | | |
| | | | | | | | |
| | | | | | | | Adults depart a month before immatures |
| | | | | | | | |
| | | | | | | | Immatures strongly dispersive |
| | | | | | | | |
| | | | | | | | |
| 8 | From 2 | Mainly Scandinavia | | | | Occasional | |
| | | | | | Yes | | |
| | | | | | | | |
| | | | | | | | |
| | | | Yes | | | | |
| | | | | | | | Mainly 8 to mid 9 and late 4 and early 5 |
| | | | | | | | Mainly 8 to mid 10 and 3 to 5 |
| | | | | | | | |
| | | | | | | Occasional | |
| | | | | | | | |
| | | | | | | | |
| 10 | 1 to 3 | North and East Europe | | | | | |
| 10 and 11 | 3 | Mainly Scandinavia | | | | | |
| | | | | | | | |
| | | | | | | | |
| | | | | | | | |
| | | | | | | | |
| and 10 | 3 and 4 | Scandinavia | | | Yes | | |
| 0 | mid 3 to early 4 | Scandinavia | | | | | |
| 0 | 4 | Central European mountains | | | | | |
| | | | | | | | |
| | | | Yes | | Yes | | |
| | | | | | Yes | | European race 'White Wagtail' is passage migrant, mainly mid 3 to 4 and 8 to 10 |
| 0 to 12 | 3 to 4 | Scandinavia and Russia | | | | | |
| | | | | | | | |
| | | | | | | | |
| | | | | | | | |
| to 11 | | | | | | | |
| | | | | | | | |
| ate 9 to 11 | 3 and 4 | North and ?Central Europe | | | | | For passage times *see* Winter Season section |
| | | | | | | | |
| | | | | | | | |
| | | | | | | | |
| | | | | | | | Greenland Wheatears peak late April to mid May |
| | | | | | | | |
| 0 to 11 | 3 | Northern Europe | No | | | | |
| 0 to 12 | late 3 and 4 | Northern Europe | Few | | | | |
| id 9 to early 11 | 3 to early 5 | Low Countries | | | | | |
| te 9 to 11 | late 2 to early 5 | Iceland and Scandinavia | | | | | |

| | Resident | Summer Visitor | Winter Visitor | Mainly Passage | Migrates within Britain | Summer Season | | | |
|---|---|---|---|---|---|---|---|---|---|
| | | | | | | Initial arrival | Main arrival | Departure | Winter destination |
| Mistle Thrush | XXXXXX | | | | A little | | | | |
| Cetti's Warbler | XXXXX | X | | | | | 4 | 9 | Continental Europe? |
| Grasshopper Warbler | | XXXXXX | | | | mid 4 | late 4 to 5 | 7 and 8 | West Africa? |
| Savi's Warbler | | XXXXXX | | | | early 4 | late 4 | 8 | Sub-Saharan Africa |
| Sedge Warbler | | XXXXXX | | | | early 4 | mid to late 4 and 5 | early 8 to late 9 | Sub-Saharan Africa |
| Marsh Warbler | | XXXXXX | | | | late 5 | early 6 | 8 to 9 | South-east Africa |
| Reed Warbler | | XXXXXX | | | | late 4 | 5 | 8 and 9 | Central Africa |
| Icterine Warbler | | | | XXXXXX | | | | | |
| Dartford Warbler | XXXXXX | | | | | | | | |
| Barred Warbler | | | | XXXXXX | | | | | |
| Lesser Whitethroat | | XXXXXX | | | | mid 4 | late 4 to early 5 | 8 and early 9 | Ethiopia and Sudan |
| Whitethroat | | XXXXXX | | | | early 4 | mid 4 to mid-5 | 8 to mid 9 | Sub-Saharan Africa |
| Garden Warbler | | XXXXXX | | | | mid 4 | late 4 and 5 | mid 7 to 9 | Central and Southern Africa |
| Blackcap | | XXXXX | X | | | early 4 | 4 and 5 | 7 to 9 | Mediterranean and West Africa |
| Wood Warbler | | XXXXXX | | | | late 4 | early 5 | 7 and 8 | Central Africa |
| Chiffchaff | | XXXXX | X | | | early 3 | late 3 to 4 | 9 and 10 | Mainly Iberia and North Africa |
| Willow Warbler | | XXXXXX | | | | late 3 | 4 to 5 | end 7 to end 9 | Central and Southern Africa |
| Goldcrest | XXXXX | X | | | | | | | |
| Firecrest | | XXX | XXX | | | early 4 | late 4 to early 5 | late 7 to 8 | SW Europe? |
| Spotted Flycatcher | | XXXXXX | | | | early 5 | mid to late 5 | 7 to 9 | Sub-Saharan Africa |
| Pied Flycatcher | | XXXXXX | | | | mid 4 | 5 and 6 | 8 and 9 | West Africa |
| Bearded Tit | XXXXXX | | | | Yes | | | | |
| Long-tailed Tit | XXXXXX | | | | | | | | |
| Marsh Tit | XXXXXX | | | | | | | | |
| Willow Tit | XXXXXX | | | | | | | | |
| Crested Tit | XXXXXX | | | | | | | | |
| Coal Tit | XXXXXX | | | | | | | | |
| Blue Tit | XXXXXX | | | | | | | | |
| Great Tit | XXXXXX | | | | | | | | |
| Nuthatch | XXXXXX | | | | | | | | |
| Treecreeper | XXXXXX | | | | | | | | |
| Golden Oriole | | XXXXXX | | | | late 4 | 5 | 8 | Central and Southern Africa |
| Red-backed Shrike | | | | XXXXXX | | | | | |
| Great Grey Shrike | | | | XXXXXX | | | | | |
| Jay | XXXXXX | | | | | | | | |
| Magpie | XXXXXX | | | | | | | | |
| Chough | XXXXXX | | | | | | | | |
| Jackdaw | XXXXX | | X | | | | | | |
| Rook | XXXXXX | | | | | | | | |
| Carrion Crow | XXXXX | | X | | | | | | |
| Raven | XXXXXX | | | | | | | | |
| Starling | XXXXX | | X | | | | | | |
| House Sparrow | XXXXXX | | | | | | | | |
| Tree Sparrow | XXXXXX | | | | | | | | |
| Chaffinch | XXXX | | XX | | | | | | |
| Brambling | | | XXXXXX | | | | | | |
| Greenfinch | XXXXX | | X | | Yes | | | | |
| Goldfinch | XXXX | | | | | 2 | 4 to 5 | 9 to 11 | France, Spain, Belgium |
| Siskin | XXX | X | XX | | | 3 | 4 | mid 9 to 10 | Iberia |
| Linnet | XXXX | XX | | | | early 3 | mid 3 to 4 | 8 and 9 | France, Spain, Morocco |
| Twite | XXXXX | XX | | | Yes | | | | |
| Lesser Redpoll | XXXXX | | | | | 3 | 4 | 10 | France |
| Crossbill | XXXXXX | | | | Yes | | | | |
| Scottish Crossbill | XXXXXX | | | | | | | | |
| Bullfinch | XXXXXX | | | | | | | | |
| Hawfinch | XXXXX | | X | | | | | | |
| Snow Bunting | XX | | XXXX | | | | | | |
| Yellowhammer | XXXXXX | | | | | | | | |
| Cirl Bunting | XXXXXX | | | | | | | | |
| Reed Bunting | XXXXXX | | | | Yes | | | | |
| Corn Bunting | XXXXXX | | | | | | | | |

Resident includes movements within Britain

| Winter Season | | | Hard-weather Movements | Moult Migration | Altitudinal Migration | Irruptions into Britain? | Notes |
|---|---|---|---|---|---|---|---|
| Main Arrival | Main Departure | Breeding destination | | | | | |
| | | | | | | | |
| | | | | | | | |
| | | | | | | | |
| | | | | | | | |
| | | | | | | | |
| | | | | | | | |
| | | | | | | | |
| | | | | | | | Mainly 8 to 10 |
| | | | | | | | |
| | | | | | | | mid 8 to late 10 |
| | | | | | | | |
| | | | | | | | |
| | | | | | | | |
| ?? | ?? | Germany | | | | | |
| | | | | | | | |
| late 10 and 11 | mid 3 to early 4 | Unknown; probably North and East Europe | | | | | |
| | | | | | | | |
| late 8 to early 11 | late 3 to early 4 | Scandinavia | | | | | |
| early 9 to 11 | late 3 to early 4 | North and East Europe | | | | | |
| | | | | | | | |
| | | | | | | | |
| | | | | | | Sometimes | |
| | | | | | | | |
| | | | | | | | |
| | | | | | | | |
| | | | | | | | |
| | | | | | | | |
| | | | | | | | |
| | | | | | | | |
| | | | | | | | |
| | | | | | | | |
| | | | | | | | |
| | | | | | | | Mainly late 5 and late 8 to 9 |
| ate 9 to 11 | 3 and 4 | Scandinavia? | | | | | |
| | | | | | | Occasional | |
| | | | | | | | |
| | | | | | | | |
| 0 and 11 | mid 2 to 4 | Northern Europe | | | Yes | | |
| | | | | | | | |
| 0 and 11 | early spring | Scandinavia* | | | | | *Hooded Crows |
| | | | | | | | |
| te 10 and early 11 | 3 to 4 | North-west Europe | | | | | |
| | | | | | | | |
| | | | | | | | |
| to 11 | 2 to 5 | Scandinavia | | | | | |
| id 9 to 10 | 3 to 4 | Scandinavia | | | | | |
| ) and 11 | 3 and 4 | Northern Europe | | | | | |
| | | | | | | | |
| id 9 and 10 | 3 to mid 4 | North-west Europe | | | | | |
| | | | | | | | |
| | | | | | | | Breeders from Pennines to North Sea coast, 10 to 3 |
| | | | | | Yes | | |
| | | | | | | Yes | Nomadic |
| | | | | | | | |
| | | | | | | Sometimes | |
| e 10 to early 11 | 4 and 5 | North-west Europe? | | | | | |
| d 10 and 11 | late 2 to 4 | Iceland, possibly Greenland | | | | | |
| | | | | | Yes | | |
| | | | | | | | |
| | | | | | | | |
| | | | | | | | |

# GLOSSARY

**Abmigration** – a return migration taking an individual to a region far removed from the starting point of its outward migration.

**Altitudinal Migration** – a seasonal movement not so much of latitude as of altitude, birds moving less between places than between heights.

**Anticyclone** – an area of high atmospheric pressure, usually associated with light winds and clear skies.

**Arched Migration** – a gradual shift of direction during the migratory journey.

**Breeding Grounds** – a broad term for the geographical range over which a species normally breeds.

**Breeding Site Fidelity** – the act of returning to the same breeding site year after year, usually for a bird's whole lifetime.

**Broad Front Migration** – basically, the migration of a species over a broad swathe of land, usually in a wide series of many parallel streams like the lanes of a motorway.

**Circannual Rhythm** – the internal programme that determines the physiological timetable of birds – when they should breed, moult, migrate etc.

**Depression** – an area of low atmospheric pressure, usually associated with unsettled weather.

**Differential Migration** – the migration of a species in which different age and sex classes have different destinations, routes and/or times of flight.

**Drift Migration** – the movement of birds off course during their migratory flight, usually caused by strong side winds.

**Dynamic Soaring** – a type of soaring that seabirds use to harness the energy of the wind over the sea; the birds fly into the wind to be lifted upwards, then use gravity to soar down before turning into the wind again.

**Escape Movement** – a movement of a bird in response to untenable weather conditions.

**Fall** – an unusually large arrival of migrant birds, usually brought about by unusual weather conditions.

**Graded Migration** – a migration in which a bird performs a full outward journey, but returns only part of the way.

**Heligoland Trap** – a special bird-catching trap developed on the island of that name. It consists of a netted funnel with a cage at the narrow end.

**Hooked Migration** – a pronounced shift of direction during the migratory journey.

**Hyperphagia** – the act of increasing food consumption (usually its frequency) in accordance with internal physiological rhythms.

**Hyperlipogenesis** – simply understood, the rapid production of fat reserves by the liver.

**Infrasound** – sounds that are too low in pitch to be audible to us (below 20Hz).

**Irruption** – the periodic but irregular arrival of a species or population well out of normal range, usually in response to overcrowding and/or lack of food. Such birds are erupting.

**Isobar** – a line joining areas of equal barometric (atmospheric) pressure.

**Leading Lines** – geographical features that direct birds during their migration – coastlines and rivers, for instance.

**Leapfrog Migration** – migratory journeys that take some populations further than others, the former overflying the latter.

**Loop Migration** – a migration in which the outward and return journeys take different pathways, although they have the same departure/arrival points.

**Magnetic Sensitivity** – a sensitivity to magnetic field that birds possess, but we don't.

**Migrant** – in one sense, a bird species that migrates; alternatively, an individual bird in the act of migrating.

**Migratory Divide** – a divergence of initial migratory direction taken by members of the same species living in different geographical areas, in order to avoid a common barrier such as a mountain range or water body.

**Mist Net** – a fine netting used to catch birds for ringing.

**Moult Migration** – a journey undertaken to a special place with the express purpose of moulting.

**Narrow Front Migration** – movement by a whole or significant proportion of a population along barrow corridors, from which the birds seldom stray.

**Nomad** – a bird that moves in response to the availability suitable breeding conditions.

**Olfactory Gradient** – a scent trail that can be followed on the basis of how strong the smell is.

**Outward Migration** – the migration that takes a bird away from its breeding areas.

**Overshoot** – a bird that overshoots the finishing line of its migratory journey and may turn up in unexpected places.

**Partial Migration** – migration in which only certain individuals of a population or species migrate, while the rest don't.

**Passage Migrant** – a bird passing through a place, of a species or population that neither breeds nor winters there.

**Post-juvenile Dispersal** – the movement of juvenile birds away from their parents' territory and beyond once they are independent, in no particular direction, and often not very far.

**Registration Cage** – a special round cage designed to measure the directional preferences of a moving bird.

**Resident** – a bird that remains all year in the same area.

**Retarded Migration** – a migratory strategy in which a bird (usually a first-year) performs a full outward journey, but may delay its return migration for a whole year or more.

**Return Migration** – the migration that takes a bird back to its breeding areas.

**Sedentary** – describing a bird or species that doesn't travel i.e. is a non-migrant.

**Slope-soaring** – a type of soaring used by seabirds to harness the energy of wind passing over waves.

**Staging Area** – an area, usually traditional, used by a bird or population of birds in mid-migration, often for refuelling, moulting and fattening.

**Star Compass** – a mechanism that enables the bird to maintain constant direction by use of the stars.

**Sun Compass** - a mechanism that enables the bird to maintain constant direction by use of the sun and its setting point.

**Thermal** – a bubble of air heated by the sun and rising from the ground, composed of swirling updrafts that can be used by birds for lift.

**Wintering Grounds** – the broad area where a species resides during the non-breeding season (also known as Resting Areas).

**Wintering Site Fidelity** – the act of returning to the same wintering site every year after breeding.

**Wreck** – the large-scale beaching or blowing inland of birds that are usually found far out at sea. Usually occurs in response to extreme weather.

**Zugunruhe** – specifically, the migratory activity of caged birds, mainly shown by perch-hopping and wing-whirring; can be applied to the pre-programmed "migratory energy" of free-living birds.

# BIBLIOGRAPHY

Alerstam, Thomas. 1990. *Bird Migration*. Cambridge University Press.

Berthold, Peter. 2001. *Bird Migration: A General Survey* (Second Edition). Oxford University Press.

Del Hoyo, J., Elliott, A. and Sargatal, J. (edds) 1992–2001. *Handbook of the Birds of the World, Vols 1–6*. Lynx Edicions, Barcelona.

Elphick, C., Dunning, J.B. (Jr) and Sibley, D. 2001. *The Sibley Guide to Bird Life and Behaviour*. Christopher Helm, London.

Elphick, J. (ed). 1995. *The Atlas of Bird Migration*. Random House, New York.

Gibbons, D.W., Reid, J.B. and Chapman, R.A. (eds) 1993. *The New Atlas of Breeding Birds in Britain and Ireland: 1988–1991*. T & A.D. Poyser, London.

Holden, P. and Cleeves, T. 2002. *The RSPB Handbook of British Birds*. Christopher Helm, London.

Lack, P. (ed). 1986. *The Atlas of Wintering Birds in Britain and Ireland*. T & A.D. Poyser, Calton.

Perrins, C (ed). 2003. *The New Encyclopedia of Birds*. Oxford University Press.

Snow, D.W. and Perrins, C.M. (eds) 1998. *The Birds of the Western Palearctic*. Concise edition. Oxford University Press.

Wernham, C.V., Toms, M.P., Marchant, J.H., Clark, J.A., Siriwardena, G.M. and Baillie, S.R. (eds) 2002. *The Migration Atlas: Movements of the Birds of Britain and Ireland*. T & A.D. Poyser, London.

# ACKNOWLEDGEMENTS

My thanks are due to the authors of the books that explained migration to me, and opened up this exciting world for us all. Any inaccuracies or misrepresentations in this book are mine alone.

Many thanks to my wife Carolyn, daughter Emily and son Samuel, who endured the usual pitfalls of living in the same house as a writer: books everywhere, jobs left undone, not enough visits to the soft-play centre, an occasionally over-absorbed husband and father. I love you all to bits.

And thanks to Charlotte Judet, long-suffering and patient editor who has a rare gift of combining good humour and professionalism.

# PICTURE CREDITS

**Illustrations** All David Daly except: 24(t), 26, 30, 35, 39, 40, 42(b), 45(t, b), 61, 63(b), 65, 70, 75, 78(b), 81, 83(t), 85(l), 87, 95(b), 106(l,r), 107(t), 108, 112(t,b), 117, 121 Richard Allen; 82(t) Stephen Message. **Photographs**: All Windrush Photos except: front jacket Laurie Campbell and rspb-images.com; 103(t) David Cottridge; 27(b), 93(t), 94(t) Rebecca Nason; 18(b) George Reszeter; back jacket, 6(l,r), 8, 9(t,b), 15, 21, 28, 51, 58, 71, 80, 89(t), 92(b), 95(t), 96(t), 97(t), 104(b), 105(t), 118(b) Steve Young. **Maps and diagrams**: All Bill Smuts.

Thanks also to Colin Pennycuick, Tom Bradbury, Óli Einarsson, Myrfyn Owen, *Ibis: International Journal of Avian Science* and *Journal of Avian Biology* for permission to reproduce the Whooper Swan flight paths on p98:

Pennycuick, C .J., T. A. M. Bradbury, Ó. Einarsson, M. Owen 1999. Response to weather and light conditions of migratory Whooper Swans and flying height profiles, observed with the Argos Satellite System. *Ibis* 141, 434-443, and Pennycuick, C. J., T. A. M. Bradbury, Ó. Einarsson, M. Owen 1996 MigratingWhooper Swans: satellite tracks and flight performance calculations. *Journal of Avian Biology* 27, no2, 118-134.

# INDEX

Page references in *italics* denote illustration/illustration caption